Polymer Electrolyte Transport and Response

By
Ahmed Al Maghribi

List of Contents

Chapter 1

An Introduction to Energy, Energy Storage Devices and Electrolyte Systems

1.1 Introduction to Energy

All phenomena unfolding in the world are the expression of energy flowing through various forms. Energy serves as the driving force behind all actions, enabling systems to function and transform. Due to its diverse expressions, defining energy comprehensively remains a challenge. The conservation principle dictates that energy is never created or destroyed; instead, it only undergoes transformations from one form to another or transfers between states. It remains ever-present, shaping existence in continuous motion.

Energy availability is essential for supporting human well-being, driving economic growth, and alleviating poverty since energy is considered to be the main driving force for economic growth and social development. Energy production, its storage, and distribution remain fundamental necessities for sustaining industrial growth and social development.

As environmental problems intensify, the importance of energy-related discussions has expanded, highlighting the urgency of identifying sustainable solutions. A comprehensive understanding of the various types of energy is essential, not only for academic knowledge but also for informing public policy and fostering innovation in energy technologies. The concept of sustainable energy emphasizes using resources that satisfy present consumption but do not threaten long-term energy availability or the stability of the climate for generations to come. Transitioning to a sustainable energy future necessitates a nuanced understanding of the spectrum of energy types and their potential towards meeting the growing demand without compromising environmental integrity.

The processing and use of diverse energy resources, beginning from fossil fuels to renewable sources, are integral to technological advancements, economic development, and overall human well-being. For instance, the interdependent connection that exists between energy provision and infrastructure development underscores the significance of energy in facilitating essential services such as healthcare and transportation. Moreover, the ongoing transition towards sustainable energy solutions is critical for mitigating environmental challenges and enhancing energy security globally. As the demand for cleaner energy alternatives grows, innovative approaches highlight the necessity for efficient energy systems. These systems not only enhance operational capabilities but also promote the need for continuous education on energy management within organizations (Futcher et al.) [1]. Ultimately, understanding energy's role in shaping modern society is crucial for navigating future challenges and opportunities. While the Sun provides the most fundamental source of energy for Earth, additional energy is harnessed from resources like coal, fossil fuels, geothermal systems, wind, biomass, nuclear power, and numerous others.

The classification of energy into various forms is made using the following set of criteria:

- **Primary and secondary energy**
- **Conventional and non-conventional energy**
- **Commercial and non-commercial energy**
- **Renewable and non-renewable energy**

1.1.1 Primary and secondary energy

Energy resources considered "primary" are those found directly in nature and used without first being transformed. These include fossil fuels like coal, oil, and natural gas, as well as wood, nuclear fuels, solar energy, wind, tides, river flows, mountain lakes, and geothermal heat. Primary energy sources are the raw materials extracted from the environment, serving as the initial stage of energy classification for electricity generation [2]. However, these sources require significant conversion processes into secondary energy forms. Among secondary energy carriers, electricity holds a central role, as it can be obtained from a wide spectrum of primary resources. These include conventional fuels like coal, oil, natural gas, biofuels, and nuclear power, along with renewable options such as solar radiation, wind, hydropower, and geothermal energy. However, the use of renewable primary resources including solar, wind, and hydro, to generate secondary energy highlights a transition to sustainable pathways, promising lower greenhouse gas emissions and a diminished impact on the environment [3].

Since the renewable primary energy share of the energy market is gradually increasing due to the increasing integration of renewable energy and supportive government measures driving the shift toward a low-carbon economy. However, global energy use still depends largely on primary sources drawn from non-renewable options such as fossil fuels and nuclear power. Hence there is a demanding need to diversify the energy mix to mitigate climate change and ensure energy security and sustainability for future generations.

Electricity and other secondary energy sources derived from renewable primary sources such as solar, wind, hydro and geothermal energy are now getting much importance for developing and implementing sustainable, efficient, and reliable secondary energy sources that can meet the increasing global energy demands [4]. Progress toward this can be made by developing and applying advanced technologies that make secondary energy sources more efficient and sustainable. Tools such as smart grids and storage systems enable better integration of intermittent renewables, reduce energy loss, and create a more flexible and resilient energy system [5].

1.1.2 Conventional and non-conventional energy

Many natural resources exist around us—some lie beneath the Earth like fossil fuels and geothermal energy, others are on the surface such as rivers and oceans, some are above in the form of wind, while the Sun serves as a major resource beyond Earth [6].

Conventional energy sources refer to those that occur naturally on or beneath the Earth and require long periods of time to form or replenish. These are generally non-renewable, meaning that once consumed, they cannot be immediately replaced. For decades, resources such as petroleum, coal, natural gas, diesel, kerosene, and propane—primarily derived from fossil fuels—have been widely utilized, with modern technologies now enabling more efficient extraction and use. Owing to their long-standing role in meeting energy needs, these resources are classified as conventional sources of energy. Within this category, they are further divided into commercial and non-commercial types.

Non-conventional, or renewable, energy sources are those natural resources that supply energy continuously and are naturally restored for repeated use even after being consumed. Unlike conventional fuels, non-conventional sources offer a sustainable alternative because they are naturally replenished and generate far less environmental risk. With the adverse effects caused by fossil fuels on the planet, there is a noticeable acceleration in the adoption of non-conventional energy sources. Sunlight, wind energy, hydropower from flowing water, and energy extracted from ocean systems are some of the key forms of non-conventional energy sources.

1.1.3 Commercial and non-commercial energy

Energy resources that are bought and sold at a definite price in the market are referred to as commercial energy. Among these, electricity, coal, and refined petroleum products are the most significant. Commercial energy plays a central role in driving modern industrial growth, agricultural productivity, transportation, and overall economic development. In the industrialized countries, for most societies, commercial fuels form the backbone of energy use, driving industries as well as powering household activities. Examples include electricity, coal, oil, natural gas, and lignite.

Non-commercial energy comprises resources that are not exchanged through formal markets and therefore do not have a fixed monetary value. Typical examples include firewood, animal dung, and crop residues, which are generally collected locally and used directly, with rural households being the primary consumers. Because they are gathered rather than purchased, they are often described as traditional fuels and are frequently excluded from official energy accounting. Examples of non-commercial energy include firewood and agricultural waste in rural areas; solar energy applied for water heating, drying of food items, and small-scale electricity generation; animal power used for transport, threshing, irrigation water lifting, and sugarcane crushing; as well as wind energy harnessed for pumping water or generating electricity.

1.1.4 Renewable and non-renewable energy:

Renewable energy is derived from natural resources that are continuously replenished by environmental processes, like wind, solar, water, and geothermal energy. Renewable energy is environmentally friendly, reliable, and low-maintenance, and is generated from natural sources that sustain or renew themselves fast enough to match human use, ensuring long-term availability. For this reason, conventional energy sources, including coal and oil, whose formation spans millions of years—are excluded from this category. Renewables present a more enduring alternative to traditional energy sources, offering growing cost-effectiveness while causing considerably less harm to the environment than fossil fuels.

The global transition in favour of renewable energy resources (RESs) is pivotal in addressing the urgent crises posed by climate change and environmental degradation. This transition is not merely beneficial; it is essential for sustainable development and energy security. Ports, for instance, are increasingly adopting RESs to enhance their energy efficiency, reducing reliance on fossil fuels that contribute significantly to atmospheric pollution [7]. These initiatives not only promote energy sustainability but also underscore the technological advancements aimed at integrating renewable resources into existing infrastructures. Furthermore, bio energy emerges as a vital cornerstone of this transition, boasting sustainability, eco-friendliness, and the potential for low-cost raw materials compared to traditional energy sources [8]. By harnessing resources such as agricultural residues and municipal solid waste, countries can significantly lower their carbon footprints while fostering a green economy. Hence, the development and implementation of RESs are critical for forging a sustainable energy future.

Types of renewable energy:

The increasing need for cleaner energy solutions has brought considerable focus on renewable sources, with solar, wind, hydro, and geothermal being the most prominent.

a) **Solar Energy:** Solar energy is produced as a result of nuclear fusion occurring inside the Sun and reaches Earth as electromagnetic radiation. Captured by photovoltaic (PV) cells, this clean and abundant resource is converted into usable electricity for residential and industrial applications. Despite requiring extensive surface area and consistent sunlight, solar power is recognized as one of the most attractive renewable sources owing to its flexible applications and ability to be scaled effectively [9]. Like geothermal systems, it is used not only to supply heat directly but also to generate electric power. To fully realize its potential, integrating solar farms with efficient energy storage systems is essential.

b) **Wind Energy:** Wind energy, abundant in coastal zones and mountainous areas with strong, consistent airflow, is harnessed using large wind turbines that convert kinetic motion into electricity. As a highly efficient renewable source, it helps reduce harmful greenhouse gases and supports the move toward cleaner, sustainable power generation.

c) **Hydroelectric energy:** Hydropower, an early form of electricity production, utilizes the kinetic force of water in motion, converting it into practical electrical energy. Historically, it supplied a major share of global electricity and remained the leading renewable source until 2019, relying on large volumes and strong water flow to operate efficiently.

d) **Biomass:** Biomass refers to organic matter that can be burned or transformed into liquid or gaseous fuels. In power plants, it is first burned to create heat, which is then turned into mechanical energy and used to generate electricity. Beyond combustion, biomass can be refined and transformed into fuels suitable for vehicles such as **biodiesel, ethanol, and methane**, depending on application needs. Historically, wood biomass was a primary energy source in the U.S. before fossil fuels became dominant. Today, biofuels derived from crops and waste materials offer renewable alternatives for energy and transportation.

e) **Geothermal:** Geothermal resources may be applied directly for heating purposes or converted into electrical power. Geothermal energy, derived from the Earths internal heat, presents a viable alternative that can supply energy independent of weather conditions, thus ensuring energy security [10].

f) **Hydrogen energy**: Hydrogen is a clean fuel that can be produced using renewable energy sources like solar, wind, and hydropower, making it a promising option for a low-carbon future. It is generated through electrolysis, a process in which water is separated into hydrogen and oxygen using electricity derived from renewable energy sources, and is known as "renewable hydrogen" or "green hydrogen". It will serve an essential role in cutting emissions from sectors where other approaches are less viable or come with higher costs.

g) **Tidal Energy:** Tides occur because of the gravitational pull of the sun and moon acting on Earth, coupled with the planet's rotation. The rise and fall of tides create potential energy, and the flow of water creates kinetic energy. These forms of energy can be captured by tidal energy technologies and converted into electricity. Collectively, these renewable energy types not only support sustainable development but also represent a critical shift away from reliance on depleting fossil fuels.

Non-renewable energy resources are natural reserves located beneath the Earth's surface. Unlike renewable options, they are consumed far more quickly than they can be naturally replaced, often requiring millions of years to form again. Coal, oil, and natural gas are the most common examples of such non-renewable sources.

The continued dependence on fossil-based and other non-renewable sources of energy significantly shapes the global energy landscape, presenting essential advantages and profound challenges. Fossil fuels, including coal, oil, and natural gas, remain central to powering industrial processes and transportation by providing reliable energy. Nonetheless, their widespread use has profound environmental drawbacks, since the carbon released through combustion is one of the main forces driving global warming and the deterioration of natural ecosystems. As emphasized in discussions surrounding energy economics, more than 60% of the oil used in the United States comes from foreign sources, highlighting a critical vulnerability in energy security [11]. Furthermore, the push toward renewable energy is fueled by rising awareness of the environmental and health risks linked to non-renewable energy sources has made the transition essential for ensuring long-term sustainability [12].

Types of non-renewable energy

Fossil fuels and nuclear energy represent two dominant types of non-renewable energy resources, each associated with specific environmental consequences and societal implications. The reliance on fossil fuels-such as coal, oil, and natural gas-poses significant challenges, notably in terms of greenhouse gas emissions and environmental degradation. As highlighted, the dependence on these resources has been a major contributor to rising carbon dioxide levels in the atmosphere, necessitating a re-evaluation of energy consumption patterns to mitigate environmental impact [13]. Conversely, while nuclear energy offers a lower carbon footprint compared to fossil fuels, it raises concerns surrounding radioactive waste management, safety, and the potential for catastrophic failures. As stated in a study on Japan's energy landscape, the continued use of fossil fuels has been shown to correlate positively with $CO2$ emissions, emphasizing the urgent need for a transition to cleaner energy sources [14]. Thus, the future of energy production must balance the immediate benefits of these non-renewable resources with their long-term environmental consequences.

Our existing energy practices contribute significantly to environmental degradation. To avoid the worst effects of climate change and stay on track with global climate commitments, shifting towards cleaner energy sources has become a necessity rather than a choice.

The global energy framework is in the midst of a progressive shift, moving away from traditional sources toward more sustainable alternatives. This shift is creating space to explore a broader spectrum of energy possibilities. In the past, energy sources were often categorized using subjective labels—typically distinguishing between fossil fuels and renewables. Such classifications, however, tended to overlook unconventional sources like vibration energy, sound waves, and even heat generated by the human body. Likewise, residual outputs from existing systems—such as turbine waste heat or flue gas emissions—were commonly viewed as inefficiencies rather than viable energy assets. Today, a more objective and evidence-based framework has emerged, classifying energy into renewable, reserve, and capture-based categories, each grounded in solid theoretical principles. As the world strives to meet sustainable development targets, there is growing consensus that low carbon emissions should be the central criterion for evaluating energy sources and shaping economic policies.

1.2 Energy Storage Devices

1.2.1 Introduction:

Energy storage involves saving energy when it's generated so it can be used later when needed [15]. Energy storage systems (ESSs) are considered to be the most practical and efficient solutions to maintain equilibrium between energy generation and its consumption. Energy storage enhances energy security as it ensures steady power supply on demand.

Numerous energy storage methods are already applied in the energy sector, and ongoing technological progress is driving the development of additional solutions that will be indispensable to the global energy landscape of the future. From small households to large industries, energy storage systems of different types and sizes provide dependable and eco-friendly backup power, including for remote sites and data centers.

Energy storage technologies are essential to speeding up the elimination of greenhouse gases through the replacement of fossil fuels with renewable energy. Energy storage systems make it possible to capture electricity generated from renewable sources such as solar and wind, and release it when demand is highest. On a large scale, this could reduce emissions by nearly one million tons of CO_2 annually for every gigawatt of installed capacity.

1.2.2 Current issues in energy storage devices:

Maximizing the performance of intermittent renewables like wind and solar depends greatly on efficient energy storage. Unlike conventional fuels such as coal and natural gas that provide stable output, renewable energy generation is subject to fluctuations caused by climatic variations [16,17].

The variable nature of renewable resources makes it necessary to capture surplus energy during high generation periods and feed it back into the grid during demand peaks. Efficient storage solutions help smooth out these fluctuations and keep the supply stable [16,18]. A lack of efficient storage limits the ability of renewable systems to supply steady power, especially when generation is low, increasing the risk of grid imbalances [19].

Human life in the 21st century has become deeply intertwined with electrochemical energy storage (EES) technologies. Over the past few years, these devices have been key to the expansion of consumer electronics in the mass market, i.e., the "decade of the smartphone" or the "decade of digital dependency" is completely rely on EES.

(i) Energy produced from renewables, including wind and solar farms, can be captured and stored using EES devices.

(ii) Electric vehicles draw their power from EES devices, making them strong contenders against traditional cars that rely on internal combustion engines.

(iii) EES devices are used in pacemakers.

Electrochemical energy storage (EES) systems employ a wide range of rechargeable batteries, including lithium-ion, lead–acid, nickel–cadmium, nickel–metal hydride, sodium–sulphur, sodium-ion, and molten metal types. Among these, lead–acid and lithium-ion batteries are the most widely adopted, accounting for more than 81% of newly installed energy storage capacity worldwide [20]. The majority of these installations are designed for short-duration applications, typically lasting from a few minutes to a few hours, and are primarily used to provide power support.

The challenges with the EES for not able to store energy as per demand and to capture the market over energy from fossil fuels includes- low energy density, limited life time, time consuming charging-discharging, limitation of the temperatures and vibrations resulting in potentially dangerous leakages, flammable and toxic, ecological contamination, and so on [21].

Energy derived from renewable sources, particularly solar and wind are highly variable, as their generation strongly depends on changing weather patterns and seasonal variations. To enable these renewables to supply a larger share of global electricity—and eventually dominate the energy mix—there is a critical need for stationary storage technologies that are not only economical and flexible, but also environmentally sustainable and capable of operating over long durations. Unfortunately, current energy storage technologies are still inadequate to fully address this challenge. Device-based systems, including batteries, semi-flow batteries, capacitors, and flywheels, are widely recognized for their efficiency and fast response times, yet they can only provide energy storage for relatively short periods, typically no more than a few hours, making them suitable mainly for grid balancing and short-term fluctuations [22, 23]. On the other hand, material-driven energy storage methods, for example pumped hydro and compressed air storage offer the ability to deliver energy over much longer timeframes, which is essential for overcoming extended periods of low renewable output. However, their deployment is restricted by geographic and site-specific conditions, and they demand large-scale infrastructure with significant upfront capital investment to be economically viable. These limitations highlight a pressing need for the development of new storage solutions that combine the flexibility of device-based technologies with the long-duration capabilities of material-based systems, ensuring that renewable energy can reliably meet demand at all times.

The future of battery storage systems is highly encouraging, with ongoing research driving the development of advanced technologies such as solid-state and next-generation flow batteries. Solid-state designs, in particular, hold great promise by offering higher energy density, faster charging capabilities, and longer operational lifespans compared to conventional systems [24,25]. These improvements are expected to enhance both the reliability and cost-effectiveness of storage solutions, making them more suitable for large-scale renewable energy integration. At the same time, innovations in battery materials and architectures are anticipated to improve efficiency, lower production costs, and increase durability. Together, these advancements will significantly strengthen the performance of renewable energy systems, making clean energy more practical and accessible for widespread use.

1.2.3 Types of energy storage devices:

A wide range of energy storage and conversion technologies are currently available that are used in our daily life. Some of them are: Fuel cells, Electrochemical supercapacitors, Batteries, Dye-sensitized solar cells, ECDs and Solid-state sensors.

1.2.3.1 Fuel cells

A fuel cell generates electricity through an electrochemical process that converts the chemical energy of fuels such as hydrogen, natural gas, or biogas, in combination with oxygen as the oxidizing agent. Fuel cell systems are recognized as a clean, efficient, and reliable power source, operating quietly while offering significant potential for sustainable energy applications. Unlike batteries, which can only store a finite amount of energy, fuel cells do not require periodic recharging; instead, they can continuously generate electricity as long as they are supplied with fuel and an oxidizing agent. Since oxygen is abundantly available from the air, the main limitation lies in ensuring a reliable supply of hydrogen. At present, fuel cells are applied across diverse sectors—powering homes and businesses, supporting critical infrastructure such as hospitals, grocery stores, and data centers, and delivering clean energy for a wide variety of vehicles, including cars, buses, trucks, forklifts, and trains.

A fuel cell consists of three main components: an anode (negative electrode), a cathode (positive electrode), and an electrolyte membrane that enables ions—commonly hydrogen ions to move between the two electrodes. During operation, hydrogen is supplied to the anode and oxygen to the cathode. At the anode, a catalyst splits the hydrogen molecules into protons and electrons. The protons move across the electrolyte membrane, while the electrons travel through an external circuit, generating electricity and releasing some excess heat. At the cathode, these electrons and protons recombine with oxygen to produce water, which remains the only by-product of the reaction. Because they have no moving parts, fuel cells function quietly and offer exceptional reliability.

The electrolyte membrane plays a vital role in a fuel cell, as it selectively allows only specific ions to pass between the electrodes. This controlled movement is essential, since the passage of free electrons or unwanted particles would interfere with the electrochemical reaction. Although fuel cells are available in different types and designs, they all operate on the same fundamental principle.

Fuel cells are generally classified mainly based on the electrolyte material they employ and their start-up timing characteristics, which can vary widely from just a second in proton-exchange membrane fuel cells (PEMFCs) to nearly ten minutes in solid oxide fuel cells (SOFCs).

1.2.3.2 Electrochemical supercapacitors:

Electrochemical capacitors also called Supercapacitors (SCs) are increasingly being viewed as a promising new technology for energy storage characterized by high power densities, up to 10,000 W/kg [26] with a longer life cycle exceeding 1 million cycles [27] that can be rapidly charged or discharged seconds to minutes [28, 29], ease of maintenance, environmentally friendly nature, and high standards of reliability and safety.

Supercapacitors (SCs) possess a higher power density and faster charging capability compared to conventional capacitors. They are often used alongside batteries to provide sudden bursts of energy when required [30]. Although their energy density is lower than that of batteries, it is considerably higher than that of traditional capacitors [31, 32]. A key advantage of SCs lies in their ability to charge and discharge rapidly while retaining their capacity over long periods, in contrast to rechargeable batteries, which gradually lose efficiency due to charge–discharge fatigue [31]. Unlike batteries that depend on chemical reactions, SCs store energy electrostatically within an electric field, making them highly efficient for applications demanding quick energy delivery. Moreover, SCs harness energy from decelerating automobiles, tides, wind, etc [31].

Unlike conventional capacitors that rely on a solid dielectric, supercapacitors store energy through two distinct mechanisms: electrostatic double-layer capacitance and electrochemical pseudocapacitance [33]. Together, these processes define the overall energy storage capacity of the device.

The significance of supercapacitors has grown due to their suitability for applications that demand frequent charge–discharge cycles rather than long-term, compact energy storage. They are increasingly utilized in automobiles, buses, trains, cranes, and elevators, where their role includes regenerative braking, short-term energy buffering, and providing bursts of power when required [34].

Supercapacitors (SPs) deliver significantly higher power compared to conventional electrolytic capacitors and batteries. The most common type, known as electric double-layer capacitors (EDLCs), is typically fabricated from carbon-based materials and stores energy electrostatically. In recent years, pseudocapacitors based on metal oxides have gained greater attention due to their higher specific capacitance, longer cycle life, and relatively low cost. Unlike EDLCs, they store energy through faradaic redox reactions, making them an attractive option for advanced energy storage applications.

Like in batteries, a supercapacitor operates with an electrolyte sandwiched between two electrodes, forming its fundamental structure, as shown in Figure 1.1. The two identical electrodes involve interfacial reactions leading to an overall electrochemical process in the cell.

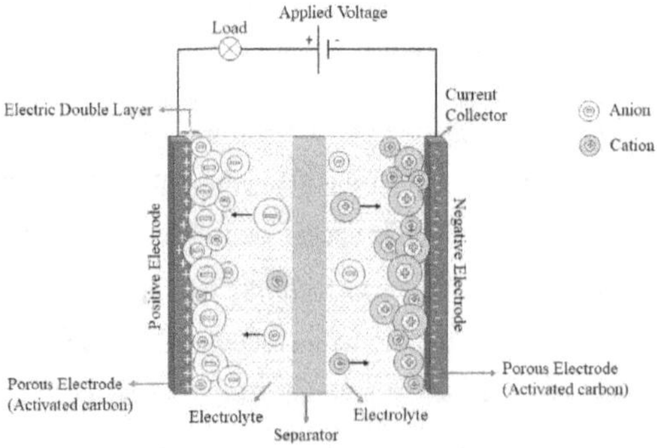

Figure 1.1 Schematic diagram of electrochemical double–layer capacitors [35,36].

Different classes of polymer electrolytes, like, SPEs, polyelectrolyte, GPEs, and composite polymer electrolytes have been used for this purpose [37, 38].

1.2.4 Batteries

1.2.4.1 Introduction

A battery works as an electrochemical system that stores energy by converting chemical energy into electricity and can reverse the process through redox reactions [39]. Batteries hold energy by storing it in chemical form and then reconverting it into electricity whenever there is a demand. It happens typically by using one or more electrochemical cells. In each cell, the anode, cathode, and electrolyte work together, allowing ions to move between electrodes and produce electricity through chemical reactions. The electrochemical processes in a battery cause electron to flow externally from the anode to the cathode. This movement constitutes an electric current, which can then be applied to perform work.

The redox reaction responses in electrochemical cells are associated with cell potential E^o, which is determined by-

$$E^o = \frac{-\Delta G^o}{zF} \qquad \text{......1.1}$$

where, ΔG^o is standard Gibbs free energy,

z - Number of electrons exchanged and

F- Faraday constant.

The theoretical cell voltage ΔE^o is obtained by subtracting the negative electrode potential $E^{0(-)}$, from the positive electrode potential, $E^{o(+)}$

$$\Delta E^0 = E^{0(+)} - E^{0(-)} \qquad \text{......1.2}$$

Electrochemical cells can be combined in series or parallel, or a combination of both to form a battery [40].

The development of modern batteries has been shaped by the pursuit of higher energy density, durability across many cycles, and improved safety standards. Ongoing innovations in electrode engineering, electrolyte formulation, and structural configuration have played a crucial role in creating more efficient and sustainable battery technologies. Modern batteries are designed to meet the specific requirements of sectors such as renewable energy integration, electric transportation, consumer electronics, and industrial power security.

Understanding battery fundamentals is crucial for developing next-generation energy storage solutions. By optimizing material compositions and electrochemical mechanisms, researchers are consistently advancing battery technologies to improve efficiency, stability, and scalability for future applications.

1.2.4.2. Classification of batteries

From a technical perspective, batteries are typically divided into two main categories based on their rechargeability: primary and secondary batteries.

Primary batteries are non-rechargeable devices designed for single use and cannot be restored once fully discharged. In primary batteries, the chemical reactions are essentially irreversible. Once the active materials in the electrodes are consumed, they cannot be restored to their original state, even if an external voltage is applied. Common examples of such batteries include zinc–carbon and alkaline types.

Secondary batteries, unlike primary batteries, are designed to be rechargeable. Their electrochemical reactions are reversible, meaning that when an external voltage is applied, the products of the discharge reaction are converted back into their original active materials. This ability allows the batteries to be used multiple times, making them more economical and sustainable compared to single-use cells. Widely used examples of secondary batteries include lithium-ion, lead–acid, and solid-state batteries, all of which play an essential role ranging from powering portable devices and electric vehicles to enabling storage solutions for renewable energy on a large scale.

1.2.4.3. Classification of battery based on ion conduction

Batteries are classified according to various parameters, one of which is ion conduction mechanism. Ion conduction plays a crucial role in determining the electrochemical performance, efficiency, and application of a battery. Based on ion conduction, batteries can be broadly categorized into **solid-state**, **liquid electrolyte**, and **polymer electrolyte batteries**.

Liquid Electrolyte Batteries: These are conventional batteries where ion conduction occurs through a liquid electrolyte. Examples include:

- Lithium-ion (Li-ion) batteries: Utilizes liquid organic electrolytes for lithium-ion transport.
- Lead-acid batteries: Uses sulfuric acid solution to facilitate ion movement.

- Nickel-Cadmium (Ni-Cd) batteries: Employs potassium hydroxide electrolyte.

Solid-State Batteries: Solid-state batteries differ from conventional designs by using a solid electrolyte instead of liquid, resulting in greater safety and improved energy density. Common examples:

- Lithium solid-state batteries: Features ceramic or sulfide-based electrolytes.
- Sodium-ion solid-state batteries: Utilizes sodium-conducting solid electrolytes.
- Superionic conductor batteries: Employs materials like NASICON (Sodium Super Ionic Conductor) for rapid ion transport.

Polymer Electrolyte Batteries: Polymer electrolyte batteries incorporate polymer-based electrolytes. Thesebatteries represent a significant innovation in electrochemical storage, offering advantages in flexibility, safety, and energy density. Unlike traditional liquid electrolyte systems, polymer electrolytes enable ion conduction through solid or gel-like polymer matrices, enhancing structural integrity and performance.

Polymer electrolytes consist of ion-conducting polymers integrated with lithium salts, sodium salts or other ionic compounds. The advantages over liquid electrolyte-based batteries include:

- Enhanced Safety: Eliminates leakage and reduces fire risks associated with volatile liquid electrolytes.
- Higher Energy Density: Enables thin, lightweight and flexible battery designs for compact applications.
- Broad Operating Temperature Range: Solid-state nature ensures stability across various temperature conditions.
- Reduced Side Reactions: Minimizes electrolyte degradation, extending battery lifespan.

The polymer electrolyte batteries are widely utilized in:

- Portable Electronics: Smartphones, tablets, and wearables benefit from their flexible form factor.
- Electric Vehicles (EVs): Solid-state polymer electrolytes promise improved longevity and safety in EV applications.

- Medical Devices: Their compact size and stability make them ideal for implants and biomedical applications.

The choice of electrolyte: liquid, solid, or polymer, determines the battery's application, whether for consumer electronics, electric vehicles, or grid storage. Ongoing advancements in solid-state and polymer electrolytes aim to enhance efficiency and sustainability in battery technologies.

In this work, the emphasis is on solid polymer electrolytes (SPEs) designed for sodium and lithium polymer batteries.

1.2.5 Development of Lithium Batteries:

Lithium batteries have revolutionized energy storage, powering everything from consumer electronics to electric vehicles. Their development has been driven by advancements in materials science, electrochemistry, and manufacturing techniques.

The concept of lithium-based batteries emerged in the 1970s, when researchers explored lithium's high electrochemical potential. Initial lithium battery designs encountered significant challenges, including dendrite growth and overall instability, which raised serious safety concerns. It was not until the 1980s that the first practical lithium-ion battery was realized, with the contribution from John B. Goodenough, M. Stanley Whittingham, and Akira Yoshino [41]. Their groundbreaking work provided the foundation for the development of modern rechargeable lithium-ion technology.

The introduction of rechargeable lithium-metal batteries during the 1980s marked a significant breakthrough in the evolution of battery technology. These early designs utilized titanium disulfide (TiS_2) as the cathode, leveraging its layered structure for lithium-ion intercalation. However, despite their promising energy density, these batteries faced significant safety concerns due to lithium metal's high reactivity, leading to thermal instability and dendrite formation.

To mitigate the risks associated with pure lithium metal, researchers explored lithium-aluminium (Li-Al) alloys as alternative anode materials. While these alloys improved safety, they exhibited lower energy density compared to pure lithium. Additionally, lithium-alloy electrodes suffered from structural degradation due to continuous swelling and shrinking of electrodes during charging and discharging, causing structural damage, resulting in fragmentation and reduced durability [42].

The limitations of lithium-metal and lithium-alloy electrodes prompted the development of graphite-based anodes. Graphite demonstrated high structural stability during lithium-ion intercalation and deintercalation, significantly reducing electrode degradation. The formation of LiC_6 during intercalation resulted in only a 10% expansion in lattice distance, making graphite a superior choice for long-term battery performance [43].

Parallel to anode improvements, researchers refined intercalation cathode materials, shifting from chalcogenides to lithium transition metal oxides such as:

- Lithium cobalt oxide ($LiCoO_2$),
- Lithium nickel oxide ($LiNiO_2$) and
- Lithium manganese oxide ($LiMnO_2$).

These materials enhanced battery efficiency, cycle life, and charge retention, paving the way for modern lithium-ion technology [44-48]. The combination of graphite anodes and lithium transition metal oxide cathodes, paired with optimized electrolytes, led to the development of lithium-ion battery technology. This innovation was first demonstrated by Murphy et al. [49] and further refined by Scrosati's research group [50]. In 1991, Sony Corporation successfully commercialized the first rechargeable lithium-ion battery, replacing lithium-metal designs with safer, high-performance alternatives.

Lithium-ion batteries have emerged as the dominant technology for powering portable electronics, electric vehicles, wearable devices, medical equipment, and large-scale energy storage systems. In operation, lithium ions move back and forth between the anode (LiC_6) and the cathode ($LiCoO_2$) through the electrolyte, while electrons travel externally through the circuit, enabling the conversion and delivery of usable energy. The electrolyte ensures ion conduction while maintaining system integrity, distinguishing intercalation-based systems from metallic lithium anodes.

The cell design of the lithium-particle battery is given underneath,

$$LiC_6(s)/\text{solvent, lithium salt}/LiCoO_2(s)$$

The transition from lithium metal to intercalation-based electrodes has led to substantial improvements in battery longevity, thermal stability, and commercial viability. However, ongoing advancements aim to address charge transfer limitations by refining electrolyte compositions and exploring solid-state and solid polymer electrolyte lithium-ion technology to enhance energy density, performance, safety, and sustainability and reduce environmental impact [51, 52].

1.2.6 Sodium Ion batteries

Sodium-ion batteries have gained attention as a promising solution for stationary energy storage on a large scale, with particular relevance to renewable sources such as solar and wind. Their viability is largely attributed to the abundant availability of sodium, which is widely present in seawater, making them a cost-effective solution compared to lithium-based systems.

To attain higher energy density, improved safety, and longer operational life, researchers are exploring alternatives to conventional hard carbon anodes. One promising approach involves metallic sodium anodes, which enable reversible plating and stripping without dendrite formation when paired with a suitable solid electrolyte. The electrolyte must possess:

- A sufficiently large energy gap to prevent unwanted reduction reactions.
- Low interfacial impedance, ensuring efficient Na^+ ion transfer between the electrode and electrolyte [53-60].

For metallic sodium anodes to function effectively, the solid electrolyte must exhibit:

- High Na^+ conductivity, exceeding 10^{-4} S/cm at operational temperatures.
- Strong wettability with metallic sodium, ensuring stable ion transport.
- Structural integrity, preventing degradation over extended charge-discharge cycles.

Current research efforts are directed toward advancing solid-state electrolytes, enhancing electrode stability, and streamlining manufacturing techniques to establish sodium-ion batteries as a practical option for grid-scale energy storage. Their strong potential to serve as an alternative to lithium-ion batteries in stationary applications continues to inspire progress in materials science and electrochemical engineering.

1.2.7 Components of a solid-state battery

As an electrochemical system, a battery consists of multiple interconnected components, each contributing critically to the processes of energy storage, transformation, and output. The efficiency, stability, and longevity of a battery are largely dependent on the properties and interactions of these components. The core components include anode, cathode, electrolyte, separator, and current collectors.

Anode: The anode serves as the negative electrode, where oxidation occurs, leading to the release of electrons into the external circuit during discharge. An ideal anode material should exhibit high stability, ease of fabrication, and minimal degradation even after numerous charge–discharge cycles. Typical anode materials consist of metals and metal alloys, such as lithium-, sodium-, and silicon-based compounds; carbon-based materials like graphite and hard carbon; as well as redox-active polymers that exhibit negative redox potentials.

Cathode: The cathode functions as the positive electrode, accepting electrons from the external circuit. To function effectively, it needs strong oxidizing characteristics, stable electrochemical behaviour, and electrolyte compatibility. Common cathode materials include Lithium transition metal oxides (e.g., $LiCoO_2$, $LiFePO_4$, $LiNiMnCoO_2$), Metal hydroxides and insertion compounds, Redox polymers with positive redox potential. The cathode's ability to maintain high voltage and capacity retention determines battery performance and energy density.

Electrolyte (Ion Transport Medium): Electrolytes form a vital part of electrochemical systems, acting as the medium that enables ion transport between the anode and cathode while blocking electron transfer. The characteristics of the electrolyte strongly affect the overall performance, efficiency, and long-term stability of energy storage devices.

Electrolytes are categorized into two types namely, solid electrolyte and liquid electrolyte. Liquid electrolytes typically used in conventional lithium-ion batteries, composed of lithium salts dissolved in organic solvents (e.g., $LiPF_6$ in ethylene carbonate mixtures). Traditional liquid electrolyte-based batteries suffer from several inherent drawbacks, including limited lifespan, slow charge–discharge rates, larger space requirements, electrode corrosion, and sensitivity to temperature fluctuations and vibrations. These issues can lead to safety risks such as leakage, flammability, and toxicity [61, 62]. In contrast, solid polymer electrolytes overcome many of these limitations. They offer higher energy density, improved scalability, and the possibility of ultra-compact designs, allowing them to store nearly twice the amount of energy as liquid electrolyte-based batteries within the same volume. The main drawback is only lower ionic conductivity and mechanical stability. The electrolytes must exhibit chemical stability, minimal reactivity with electrodes, and consistent performance across temperature variations.

1.3 Solid electrolytes

Solid electrolytes are advanced ion-conducting materials that enable ionic transport in the solid phase in electrochemical system without the use of liquid-based solvents. Unlike electronic conductors such as metals and semiconductors, which facilitate electron mobility, solid electrolytes conduct electricity exclusively through the migration of ions (cations or anions) within a structural framework. The ion conduction mechanism in solid electrolytes closely resembles that of liquid electrolytes or electrochemical solutions but occurs entirely within a solid matrix.

Their importance comes from the combination of high ionic conductivity with strong mechanical integrity and electrochemical stability. These properties make them essential for advanced technologies, particularly in the development of solid-state batteries. These materials are widely referred to as fast ionic conductors or superionic conductors due to their ability to support rapid ion transport with minimal activation barriers. This performance stems from the presence of a high density of mobile ions, ample vacant or interstitial sites, and a polarisable anion framework, which together promote effective ion hopping across energetically favourable positions.

For a material to work well as a solid electrolyte, it needs to provide excellent ionic conductivity, enabling rapid ion migration with minimal resistance to enhance charge transfer efficiency. It should possess negligible electronic conductivity to prevent electron flow, ensuring ion-driven charge transport and minimizing self-discharge. A low activation energy is essential for efficient ion diffusion, determined by crystal structure, interfacial interactions, and defect engineering. The electrolyte must contain a high concentration of mobile ions, such as Li^+, Na^+, or Mg^{2+}, to sustain ionic conduction, along with sufficient vacant sites or interstitial positions that facilitate continuous ion hopping. Lastly, a highly polarizable anion framework aids ion mobility by lowering migration barriers, further optimizing transport dynamics within the electrolyte.

Based on their structural composition and ionic transport mechanisms, Solid electrolytes are generally grouped into several categories, including crystalline solid electrolytes, glassy or amorphous solid electrolytes, composite solid electrolytes, and polymer-based solid electrolytes.

1.3.1 Crystalline Solid Electrolytes

In solid-state energy storage, crystalline solid electrolytes are of great importance due to their key functional role, offering efficient ionic transport through their rigid periodic lattice. Unlike liquid electrolytes, they enable intrinsic ion migration via vacancy diffusion, interstitial hopping, and lattice-assisted transport, delivering reliable conductivity while maintaining stability. These materials consist of a structured crystalline skeleton, where both mobile ions and immobile framework elements coexist, and enabling efficient charge transfer. The framework itself plays a decisive role in ion diffusion and conduction pathways, making these materials essential in solid-state batteries, fuel cells, and electrochemical devices.

These materials are classified as soft-framework crystals (e.g., AgI, CuI) with loosely bonded lattices for enhanced ion mobility, and hard-framework crystals (e.g., stabilized zirconia), which maintain rigid atomic arrangements for improved thermal stability. The selection of cation species, lattice flexibility, and anion polarizability governs electrochemical performance, making crystalline solid electrolytes a key component in solid-state batteries, fuel cells, and emerging electrochemical systems.

1.3.2 Glassy (Amorphous) Solid Electrolytes

Glassy, also known as amorphous, solid electrolytes are a class of non-crystalline ion-conducting materials characterized by the absence of long-range atomic order. Unlike their crystalline counterparts, which possess periodic lattice structures, glassy electrolytes exhibit a disordered network that offers isotropic conduction pathways. This disordered nature is not a limitation but an advantage as it provides a structurally flexible matrix favorable to uniform ion diffusion, eliminating grain boundaries that might otherwise hinder ionic transport.

The conduction of ions in glassy solid electrolytes is achieved through mechanisms facilitated by the random distribution of coordination sites and loosely bound ionic clusters. Mobile ions (typically Li^+, Na^+ or Ag^+) migrate through dynamic pathways formed by local rearrangements in the glass network. The absence of periodic constraints allows for lower activation energies in certain compositions, especially when the glass matrix is composed of highly polarizable anion groups such as PO_4^{3-}, BO_3^{3-}, or AsO_3^{3-}, which aid in the delocalization of charge and promotes salt dissociation.

1.3.3 Composite Solid Electrolytes

CSEs are hybrid materials that integrate inorganic solid electrolytes with polymers or the incorporation of flexible binders to resolve the challenges associated with purely ceramic systems, like brittleness and poor electrode interfacial contact. Typically, ceramic-based lithium-ion conductors like $Li_7La_3Zr_2O_{12}$ (LLZO), $Li_{1.3}Al_{0.3}Ti_{1.7}(PO_4)_3$ (LATP), or sulfide-based materials are used as the primary conductive phase. These ceramics stand out for their high ionic conductivity, close to $\sim 10^{-3}$ S/cm, combined with stable electrochemical performance, but lack mechanical flexibility and processability [63]. To overcome this, polymeric additives or ionic liquids are introduced to enhance flexibility and reduce interfacial resistance.

Recent studies have focused on fabricating CSEs with interpenetrating networks or percolated ceramic frameworks that maintain continuous ion-conduction pathways while enhancing structural integrity [64]. CSEs also show promise in suppressing lithium dendrite growth, a major concern associated with lithium-metal batteries. Despite these advancements, long-term use is still hindered by challenges including interfacial degradation and phase separation [65].

1.3.4 Polymer solid electrolytes (PSEs)

PSEs are an essential class of ion-conducting materials that provide flexibility, processability, and stable ionic transport in solid-state electrochemical systems. They offer a viable alternative to ceramic and glassy electrolytes, which, despite their high ionic conductivity, often suffer from brittleness and interfacial incompatibility. The unique ability of polymer electrolytes to support ion migration, while maintaining mechanical integrity makes them highly attractive for applications in SSBs, supercapacitors, and advanced energy storage technologies.

The ionic conduction mechanism in polymer solid electrolytes is primarily governed by ion mobility generated by the segmental motion of polymer chains, where ionic transport is facilitated through the polymer's dynamic chain movements. Unlike crystalline electrolytes that function via defect-assisted hopping mechanisms, polymer electrolytes operate through polymer-salt interactions, solvation effects, and chain flexibility. These results in improved ion dissociation, enhanced mobility, and reduced activation energy barriers, allowing for efficient charge transport even in solid-state conditions.

1.4 Classification of Polymer solid electrolytes

Polymer electrolytes continue to be an area of intense research due to their flexibility, processability, and interfacial compatibility making it a potential candidate for revolutionizing solid-state battery technology. The classification of PSEs is typically organized into the following categories:

1.4.1 Dry Solid Polymer Electrolytes (SPEs)

Solid polymer electrolytes (SPEs), often described as solvent-free systems, operate by incorporating inorganic salts into high–molecular weight polymers, enabling ion transport without the need for liquid components or plasticizing agents. In these systems, the polymer matrix itself serves as the host for dissolved salt, facilitating ion transport via complexation with functional groups present in the polymer backbone.

When an external potential is applied, ion transport occurs via coordination, de-coordination, and re-coordination within the polymer structure, facilitating charge mobility. The electrochemical cell's performance is strongly influenced by the salt's activity, impacting both charge transport mechanisms and phase potential variations. The absence of free solvent molecules reduces safety concerns such as leakage, evaporation, and flammability, making them highly attractive for solid-state battery applications.

Recent investigations into SPE dynamics have emphasized transference number, coordination strength, and ionic conductivity, elucidating fundamental charge transport mechanisms. In solid polymer electrolytes (SPEs), cation transport mainly occurs through the formation of polymer–salt complexes, assisted by the segmental motion of the polymer chains. Ions move by hopping between temporary coordination sites along these chains. To enhance ionic mobility, polymers with high dielectric constants—such as poly(ethylene oxide) (PEO) and polyvinylidene fluoride (PVDF)—are frequently employed.

The optimization of solid polymer electrolytes continues to be a focus in next-generation battery research, with advancements in polymer engineering and molecular tuning driving improvements in electrochemical stability, ion transport efficiency, and long-term device performance.

1.4.2 Gel Polymer Electrolytes (GPEs)

Gel polymer electrolytes are an intermediate between solid and liquid electrolytes, consisting of a polymer matrix swollen with an ionic liquid or a plasticizer. GPE retain a liquid phase within a polymer matrix, facilitating ion transport. The gel structure enhances ionic conductivity by enabling solvent retention and optimizing ion migration pathways. Amorphous regions within the polymer absorb more solvent than crystalline domains, improving ion diffusion primarily diffusion-controlled process. Solvent selection, particularly based on its dielectric constant, significantly impacts ionic conductivity by influencing salt dissociation and ion mobility. Charge percolation occurs in ordered polymer structures, but a higher concentration of amorphous domains enhances conductivity.

Among gel polymer electrolytes, PEO remains a widely studied material due to its compatibility with lithium electrodes. However, its plasticized form compromises mechanical integrity. To counteract this, poly(vinylidene fluoride) (PVDF) is often blended with PEO, reinforcing mechanical strength while preserving electrochemical properties [66].

According to recent literature, these materials are increasingly being explored for use in flexible energy storage devices, most notably lithium-ion batteries, where they mitigate leakage and improve electrochemical performance [R]. Innovations in GPEs involve ionic liquid incorporation, polymer blending, and nanostructuring, enhancing ion transport and thermal stability [67]. Their tunable properties make them ideal for advancements in solid-state energy storage, ensuring improved electrochemical performance and long-term device reliability.

1.4.3 Plasticized Polymer Electrolytes (PPEs):

PPEs are a subclass of SPEs designed by adding low molecular weight plasticizers such as ethylene carbonate (EC), propylene carbonate (PC), polyethylene glycol (PEG), or dimethyl sulfoxide (DMSO) into a polymer matrix. This modification improves ionic conductivity by reducing the crystallinity of the host polymer, enhancing segmental chain mobility and thereby facilitating more efficient lithium-ion transport. Commonly formulated using polar polymers like poly(ethylene oxide) (PEO), poly(vinyl chloride) (PVC), or poly(vinylidene fluoride-co-hexafluoropropylene) (PVDF-HFP), PPEs strike a balance between mechanical flexibility and electrochemical functionality, making them well-suited for integration into lithium-ion and lithium-metal battery technologies. At ambient conditions, their ionic conductivity can reach values near 10^{-4} S/cm, significantly exceeding that of traditional dry solid polymer electrolytes.

However, excessive plasticizer content can compromise mechanical stability and increase the risk of leakage or flammability. To address these limitations, recent research has explored dual-plasticizer systems and ionic liquid incorporation to simultaneously boost conductivity and thermal resilience. For example, Mahendran et al. (2023) reported a PVC-based electrolyte enhanced with EC and dioctyl phthalate (DOP), achieving improved interfacial stability [68]. Likewise, Zhao et al. (2022) utilized deep eutectic solvents within PEO matrices to elevate electrochemical robustness [69]. These advances highlight the potential of PPEs in flexible and low-temperature energy storage applications, though ongoing efforts are needed to optimize their mechanical integrity and long-term reliability.

1.4.4 Composite Solid Polymer Electrolytes (CSPEs):

CSPEs form a subclass of SPEs in which an ion-conductive or semi-conductive polymer matrix is combined with finely dispersed inorganic fillers for achieving improvements in the combined electrochemical, thermal, and mechanical properties. Base polymers such as PEO, PVDF-HFP, PMMA, or PAN act as host matrices for lithium salts including LiTFSI, $LiClO_4$, and $LiPF_6$. Within the amorphous regions of the polymer, these salts dissociate, releasing lithium ions that migrate through the polymer network to facilitate conduction. However, the semi-crystalline structure of many polymers limits their ionic conductivity at room temperature, making them less suitable for high-performance battery applications.

To overcome this limitation, CSPEs incorporate inert or active inorganic fillers. Inert fillers, such as Al_2O_3, SiO_2, and TiO_2, do not contribute directly to lithium-ion conduction but help suppress polymer crystallinity and enhance mechanical properties. On the other hand, active fillers like LLZO, LATP, or LAGP not only improve the polymer-filler interface but also create fast ionic transport channels within the matrix. This dual contribution results in increased amorphous content, enhanced segmental motion of the polymer chains, and improved overall lithium-ion mobility. CSPEs typically exhibit ionic conductivities in the range of 10^{-4} to 10^{-5} S/cm at room temperature, and their flexible nature ensures better interfacial contact with electrodes, especially lithium metal anodes. Moreover, CSPEs can be fabricated using scalable techniques like solution casting or hot pressing, offering practical advantages for device integration. Their application is well-suited for flexible energy storage devices, thin-film batteries, and hybrid battery systems.

1.4.5 Blend polymer electrolyte

Blend polymer electrolytes are a specific type of SPEs formed by combining two or more polymers to achieve collaborative enhancements in electrochemical and mechanical properties. The blending approach influences the unique characteristics of each polymer, optimizing ionic conductivity, structural flexibility, and interfacial stability. Typically, blend polymer electrolytes incorporate a high-ionic-conductivity polymer and a mechanically stable polymer, providing a well-balanced performance profile that makes them suitable for energy storage systems, including lithium-ion batteries, supercapacitors, and fuel cells.

One of the key advantages of blend polymer electrolytes is the improved ionic transport pathways, which arise from optimized polymer chain mobility and enhanced free volume. The presence of multiple polymer domains facilitates ion hopping mechanisms, reducing activation energy for ionic migration. Additionally, blending polymers with complementary dielectric properties enhances salt dissociation, contributing to higher charge carrier density and improved electrolyte functionality.

Mechanical durability is another critical advantage. By incorporating a high-strength polymer, blend polymer electrolytes exhibit greater dimensional stability, reducing the risk of degradation under electrochemical cycling. This property is particularly beneficial in flexible and wearable battery technologies, where mechanical resilience is essential.

Interfacial stability is significantly enhanced in blend systems, reducing unwanted side reactions between the electrolyte and electrodes. This is especially relevant for solid-state batteries, where interfacial compatibility dictates long-term performance. Additionally, blend polymer electrolytes allow for optimized electrochemical properties, enabling fine-tuned formulations that balance conductivity, flexibility, and thermal stability.

Overall, blend polymer electrolytes represent a promising approach to overcoming limitations associated with single-polymer electrolytes. Their ability to integrate the best attributes of multiple polymers makes them suitable for a wide range of next-generation energy storage systems, where they contribute to safer, more efficient, and commercially viable batteries.

1.5 Literature Review

1.5.1 Polyvinyl Alcohol (PVA)

Polyvinyl alcohol (PVA) is a polymer that exhibits semi-crystalline characteristics with a unique molecular structure that makes it highly suitable for solid polymer electrolytes (SPEs). Its backbone consists of hydroxyl (-OH) functional groups, which play a vital role in ionic transport and polymer-salt complexation. The presence of these hydroxyl groups enables strong hydrogen bonding, facilitating ion coordination and enhancing electrolyte stability.

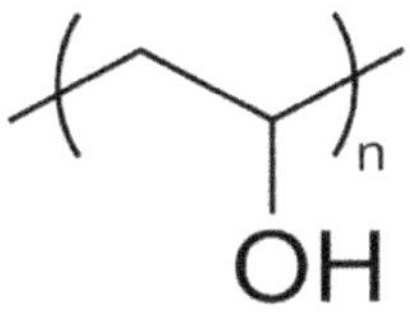

Figure 1.2 Structure for polymer Poly [vinyl alcohol] (PVA).

In this polymer, the amorphous sections allow ions to migrate, while the crystalline parts maintain mechanical integrity. The balance between these phases is critical for optimizing electrolyte performance. Additionally, PVA's hydrophilic nature allows it to retain moisture, which can further help in ionic transport when used in hybrid electrolyte systems.

The hydroxyl groups in PVA form intermolecular hydrogen bonds, contributing to its mechanical robustness. When complexed with conducting salts, such as lithium-based compounds, PVA undergoes ion-dipole interactions, where cations coordinate with oxygen atoms in the polymer matrix. This interaction facilitates ion hopping mechanisms, improving charge mobility across the electrolyte.

Recent studies have explored crosslinking strategies to enhance PVA's electrochemical stability, where covalent bonding between polymer chains reduces solubility while maintaining ionic conductivity. Additionally, blending PVA with polymers like poly(vinyl pyrrolidone) (PVP) has been shown to improve dielectric properties and optimize charge transport efficiency. PVA is known for allowing both segmental motion and molecular rotation, which play a key role in the relaxation process that occurs when hydrogen bonds form between polymers.

Research highlights its ability to coordinate with conducting salts, forming polymer–salt complexes that enhance ionic conductivity. Findings from previous studies on the conductivity of PVA polymer electrolytes with different salts applied in both energy storage and medical fields are compiled in Table 1.1.

Table 1.1 Literature survey for polymer poly [vinyl alcohol] (PVA)

S. No.	Polymer	Conductivity (S/cm)	References
1	PVA-$NaNO_3$	6.69×10^{-7}	[70]
2	PVA–$LiCF_3SO_3$	7×10^{-4}	[71]
3	PVA/ZnO	4.8×10^{-5}	[72]
4	PVA-NH_4X (X = F, Cl, Br, I)	6.9×10^{-6}	[73]
		1.0×10^{-5}	[74]
		7.68×10^{-4}	[75]
		2.5×10^{-3}	[74]
5	PVA - NH4SCN	$2 \times 58 \times 10^{-3}$	[76]
6	PVA/TiO2	5.6×10^{-3}	[77]
7	PVA/ LiFePO4	3.12×10^{-5}	[78]

1.5.2 Poly [vinyl pyrrolidone] (PVP)

PVP is known as an amorphous polymer with excellent water solubility characterized by a linear carbon backbone made from the monomer *N*-vinylpyrrolidone. The pyrrolidone ring in each repeating unit introduces a highly polar carbonyl group, which plays a pivotal role in coordinating with ionic species. This structural feature makes PVP an excellent candidate for solid polymer electrolytes (SPEs).

The amorphous nature of PVP contributes to enhanced ionic conductivity by providing greater free volume and segmental mobility, which facilitate ion transport. Unlike semi-crystalline polymers, PVP lacks rigid crystalline domains, allowing for more uniform ion migration pathways. However, PVP faces conductivity challenges because of its rigid pyrrole group, which raises the transition temperature (Tg). Due to its Lewis base nature, the tertiary amide carbonyl group in PVP can form bonds with various inorganic salts. Its relatively high glass transition temperature (Tg), attributed to the rigidity of the pyrrolidone ring, ensures thermal stability while maintaining sufficient chain flexibility for ion conduction at ambient conditions.

From a bonding perspective, due to its Lewis base nature, the tertiary amide carbonyl group in the pyrrolidone ring acts as a coordination site for cations such as Li^+, Na^+, or NH_4^+. These ion–

dipole interactions are fundamental to the formation of polymer–salt complexes, which are essential for ionic conduction in SPEs. Additionally, PVP can engage in hydrogen bonding with other polymers (e.g., PVA) or additives, enhancing miscibility and mechanical integrity in blend systems.

Recent studies have demonstrated that PVP-based electrolytes, especially when blended with polymers like PAN or PVA and doped with salts such as ammonium thiocyanate or sodium nitrate, exhibit improved conductivity and electrochemical performance due to the synergistic effects of structural flexibility and strong ionic coordination.

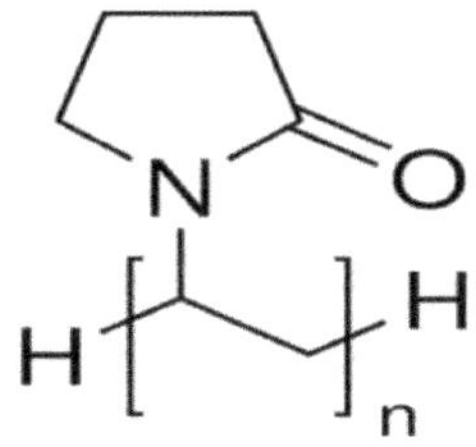

Figure 1.3 Structure for polymer Poly [vinylpyrrolidone] (PVP).

The literature review of the conductivity of PVP as Polymer electrolyte with different salts for battery applications are shown in Table 1.2

Table 1.2 Literature review of polymer PVP

S.No.	Polymer	Conductivity (S/cm)	References
1	PVP:NH_4Br	1.06×10^{-3}	[79]
2	PVP:$KBrO_4$	1.03×10^{-5}	[80]
3	PVP:$KClO_4$	9.1×10^{-6}	[81]
4	PVP:NH_4SCN	1.7×10^{-4}	[82]
5	PVP-NH_4Cl	2.51×10^{-5}	[83]
6	PVP:$MgSO_4$.7H2O	1.21×10^{-4}	[84]
7	PVP-KIO_4-TiO_2	6.665×10^{-4}	[85]
8	PVP:$MgCl_2.6H_2O$:Al_2O_3	$1.22\text{x}10^{-6}$	[86]
9	PVP-EG-Clay	2.97×10^{-3}	[87]
10	PVP-PEG-SA	8.23×10^{-7}	[88]

1.5.3 Poly (vinylidene fluoride-co-hexafluoropropylene) (PVDF-HFP)

PVDF-HFP is a semi-crystalline copolymer known for its exceptional electrochemical stability, mechanical robustness, and compatibility with lithium salts, making it a widely used host matrix in solid polymer electrolytes (SPEs). Structurally, PVDF-HFP consists of repeating units of vinylidene fluoride (VDF) and hexafluoropropylene (HFP), where the incorporation of HFP disrupts the crystallinity of PVDF, increasing the amorphous content and thereby enhancing ionic conductivity.

The crystalline domains, primarily in the non-polar α-phase of PVDF, provide mechanical strength, while the amorphous regions-enriched by HFP—facilitate ion transport by offering greater segmental mobility and free volume. This dual-phase morphology enables a balance between structural integrity and electrochemical performance. X-ray diffraction and FTIR studies confirm that the degree of crystallinity and the distribution of isomeric sequences in the amorphous phase can be tuned by processing conditions, such as solvent choice and crystallization temperature.

From a bonding perspective, PVDF-HFP exhibits strong dipole–dipole interactions due to the high electronegativity of fluorine atoms. These interactions contribute to its high dielectric constant, which aids in the dissociation of lithium salts and stabilization of free ions. The fluorinated backbone also provides chemical inertness and resistance to oxidative degradation, crucial for long-term cycling in batteries.

Moreover, the carbon–fluorine (C–F) bonds in PVDF-HFP are highly polar and thermally stable, contributing to its excellent dielectric and thermal properties. When complexed with lithium salts, cations interact with the electron-rich fluorine atoms and the polymer backbone, enabling ion hopping through coordination–de-coordination mechanisms.

Thus, the unique combination of semi-crystalline morphology, high dielectric constant, and fluorine-rich backbone in PVDF-HFP makes it a versatile and high-performance matrix for solid polymer electrolytes.

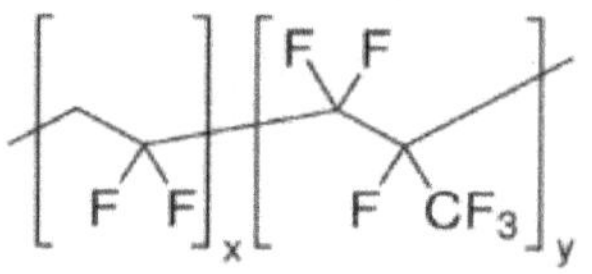

Figure 1.4 Structure for polymer PVDF-HFP.

The literature review of the conductivity of PVDF-HFPas Polymer electrolyte with different salts for battery applications are shown in Table 1.3

Table 1.3 Literature review of polymer PVDF-HFP

S. No.	Polymer	Conductivity (S/cm)	References
1	PVDF-HFP/$LiClO_4$	4.88×10^{-4}	[89]
2	PVDF-HFP + THF/acetone + $LiClO_4$ in EC/DEC	1.06×10^{-3}	[90]
3	PVDF-HFP/PANI	1.04×10^{-3}	[91]
4	PVDF-HFP/PANI/GO	6.64×10^{-4}	[91]
5	PVDF-HFP / LiTFSI (6:4 by weight; no plasticizer)	$\sim 3.28 \times 10^{-5}$	[92]
6	PVDF-HFP / LiTFSI / SN (with 20% succinonitrile plasticizer)	1.18×10^{-4}	[92]
7	PVDF-HFP / LiTFSI / SN / LLZTO (composite electrolyte)	4.05×10^{-4}	[92]

1.5.4 PVA: PVP blend polymer electrolyte

Polyvinyl alcohol (PVA) is a polyhydroxy polymer widely used in practical applications due to its simple preparation process and biodegradability [93]. In addition, PVA exhibits excellent film-forming ability, hydrophilic behavior, and a high concentration of reactive functional groups, which make it well-suited for cross-linking through irradiation, chemical, or thermal methods [94]. Polyvinyl pyrrolidone (PVP), on the other hand, acts as a stabilizer and hydrophobicizing agent, significantly enhancing the oxidative and chemical stability of membranes [95]. With these complementary properties, researchers have explored various compositions of PVA/PVP blend solid polymer electrolytes (SPEs) incorporating different salts.

Literature survey based on PVA blend with PVP with different salts as dopants is shown in Table1.4.

Table 1.4 Literature review of blend polymer PVA/PVP

S.No.	Polymer	Conductivity (S/cm)	References
1	PVA-PVP-Li2CO3	1.15 ×10-5	[96]
2	PVA-PVP-LiNO3	6.828× 10-4	[97]
3	PVA-PVP-KOH	2.2 × 10-7	[98]
4	PVA-PVP-CH3COONH4	8.12 × 10-5	[99]
5	PVA-PVP-LiClO4	10-4	[100]

1.5.5 PVDF-HFP: PVP blend polymer electrolyte:

The PVDF-HFP:PVP blend polymer electrolyte integrates the robust mechanical strength and electrochemical durability of PVDF-HFP with the highly amorphous, ion-conductive characteristics of PVP. This synergistic combination promotes efficient salt dissociation and facilitates rapid ion transport, while also imparting structural flexibility. As a result, the blend exhibits properties well-suited for use in advanced solid-state lithium-ion battery systems.

Literature survey based on PVDF-HFP: PVP polymer blend with different salts as dopants are shown in Table1.5.

Table 1.5 Literature review of blend polymer PVDF-HFP: PVP

S. No.	Polymer	Conductivity (S/cm)	References
1	PVDF-HFP:PVP:LiBr	1.13×10^{-6}	[101]
2	PVDF-HFP / PVP / EC-DMC / $LiBF_4$	$\sim 4 \times 10^{-4}$	[102]
3	PVDF-HFP/(PC + DEC)/ LiClO4	7.5×10^{-3}	[103]
4	PVDF/(PC + DEC)/ LiClO4	1.3×10^{-3}	[103]

1.6 Aims and Objectives

The main goal of this research work is to design, synthesize, and characterize efficient solid polymer electrolytes (SPEs) with suitable lithium and sodium salts, with the objective of understanding their structural, electrical, and relaxation behaviour, and to establish optimized compositions for efficient ion transport for their application in SSBs. This objective is addressed through the following specific aims:

a) Formulation and Development of SPEs: To formulate and develop suitable SPE systems by selecting appropriate copolymer host matrices and incorporating inorganic salts, with the dual aim of achieving high superior ionic conductivity while maintaining both flexibility and thermal robustness.

b) Optimization of Polymer–Salt Interactions: To systematically vary polymer compositions and salt concentrations in order to enhance ion dissociation, improve charge carrier mobility, and minimize ion–ion association that restricts conductivity.

c) Structural and Functional Validation: To validate electrolyte formation using structural and spectroscopic techniques, including X-ray diffraction (XRD) to evaluate phase composition and amorphicity, and Fourier transform infrared spectroscopy (FTIR) to confirm functional groups and polymer–salt interactions.

d) Electrochemical and Impedance Characterization: To conduct comprehensive electrochemical impedance spectroscopy (EIS) studies on the prepared SPEs to determine ionic conductivity, activation energy, and dielectric relaxation characteristics across different compositions and temperature conditions.

e) Investigation of Ion Transport Mechanisms: To examine ion transport dynamics by correlating polymer architecture, salt concentration, and thermal conditions with ionic conductivity, thereby establishing the fundamental conduction mechanisms in the SPE systems.

1.7 Scope of the Thesis

The present thesis is directed toward the systematic exploration of polymer blend electrolytes as alternative solid-state ion-conducting materials for solid-state batteries (SSBs). The work is structured into two distinct but complementary investigations that collectively aim to establish design principles for high-performance, flexible, and environmentally sustainable solid polymer electrolytes.

In Chapter 3 and 4, PVDF-HFP/PVP-based electrolytes incorporated with lithium salts (LiI and $LiNO_3$) were examined to elucidate the impact of fluoropolymer–pyrrolidone blending on structural modification, ionic transport, and dielectric relaxation. These studies focused on understanding how the introduction of lithium salts influences amorphicity, polymer–salt interactions, and the resulting enhancement of ionic conduction pathways.

In Chapter 5, the research transitions to sodium-ion systems, addressing the need for cost-effective and abundant alternatives to lithium. For this purpose, PVA/PVP blends doped with sodium sulfate (Na_2SO_4) were prepared and investigated using electrochemical impedance spectroscopy and modulus analysis. The scope here was to determine the role of salt concentration in governing bulk resistance, relaxation dynamics, and conductivity, with emphasis on identifying the optimal composition capable of supporting efficient Na^+ transport.

By integrating insights from both lithium- and sodium-based polymer electrolytes, the thesis defines a broader framework for optimizing polymer blend systems toward specific energy storage chemistries. The combined outcomes highlight the interplay of polymer selection, salt type, and ion–polymer interactions as critical factors in optimizing solid polymer electrolytes for sustainable battery applications.

1.8 References:

[1] Futcher, J.; Mills, G.; Emmanuel, R. Interdependent energy relationships between buildings at the street scale. *Building Research & Information* **2018**, *46*, 829-844

[2] Khare, V.; Khare, C.; Nema, S.; Baredar, P. *Introduction to Energy Sources: Tidal Energy Systems*; Elsevier: Amsterdam, The Netherlands, 2019; pp. 1–39, ISBN 9780128148815.

[3] Ang, T.Z.; Salem, M.; Kamarol, M.; Das, H.S.; Nazari, M.A.; Prabaharan, N. A comprehensive study of renewable energy sources: Classifications, challenges and suggestions. *Energy Strateg. Rev.* **2022**, *43*, 100939.

[4] Strielkowski, W.; Civín, L.; Tarkhanova, E.; Tvaronavičienė, M.; Petrenko, Y. Renewable energy in the sustainable development of electrical power sector: A review. *Energies* **2021**, *14*, 8240.

[5] Malik, F.H.; Hussain, G.A.; Alsmadi, Y.M.S.; Haider, Z.M.; Mansoor, W.; Lehtonen, M. Integrating energy storage technologies with renewable energy sources: A pathway toward sustainable power grids. *Sustainability* **2025**, *17*, 4097.

[6] Al Mubarak, F.; Rezaee, R.; Wood, D.A. Economic, societal, and environmental impacts of available energy sources: A review. *Eng.* **2024**, *5*, 1232–1265.

[7] Sadiq, M.; Ali, S.W.; Terriche, Y.; Mutarraf, M.U.; Hassan, M.A.; Hamid, K.; Ali, Z.; Sze, J.Y.; Su, C.L.; Guerrero, J.M. Future greener seaports: A review of new infrastructure, challenges, and energy efficiency measures. *IEEE Access* **2021**, *9*, 75568–75587.

[8] Abdellatief, T.M.M.; Abideen, Z.; Akhter, P.; Ghenai, C.; Hussain, M.; Inayat, A.; Jamil, F.; Shanableh, A. Valorization of waste biomass to biofuels for power production and transportation in optimized way: A comprehensive review. *Wiley Interdiscip. Rev. Energy Environ.* **2024**, *13*, e502.

[9] Asiata, I.; Hakeem, O.; Mubarak, L.; Yekinni, S. Solar photovoltaic energy system. *IntechOpen* **2023**.

[10] Lin, J. Techno-economic assessment of hydrokinetic turbine (HKT) application in Malaysia. *Energies* **2022**, *15*, 1234.

[11] Halcrow, S.D. *Green Energy for the Battlefield*; Master's Thesis, Naval Postgraduate School: Monterey, CA, USA, 2007.

[12] Tohma, M. *Renewable Energy Consumption and Economic Growth in European Union Countries*; Master's Thesis, Czech Technical University in Prague, Computing and Information Centre: Prague, Czech Republic, **2021**.

[13] Elliott, D. *Sustainable Energy: Choices, Problems and Opportunities*; Royal Society of Chemistry (RSC): Cambridge, UK, **2003**.

[14] Abban, A. Impact of Energy on the Environment: The Case of Japan; *Российские регионы в фокусе перемен. —Ч. 1.—Екатеринбург* **2016**, *1*, 433-442.

[15] Mitali, J.; Dhinakaran, S.; Mohamad, A.A. Energy storage systems: A review. *Energy Storage Saving* **2022**, *1*, 166–216.

[16] Conde, H.J.C.; Demition, C.M.; Honra, J. Storage Is the New Black: A Review of Energy Storage System Applications to Resolve Intermittency in Renewable Energy Systems. *Energies* **2025**, *18*, 354.

[17] Ejuh Che, E.; Roland Abeng, K.; Iweh, C.D.; Tsekouras, G.J.; Fopah-Lele, A. The Impact of Integrating Variable Renewable Energy Sources into Grid-Connected Power Systems: Challenges, Mitigation Strategies, and Prospects. *Energies* **2025**, *18*, 689.

[18] Ghose, B. Integrating Energy Storage Systems with Utility-Scale Photovoltaic (PV) Installations in the Grid. *Int. J. Eng. Sci. Math.* **2024**, *13* (10), 31–45.

[19] Su, H.-I.; El Gamal, A. Modeling and Analysis of the Role of Energy Storage for Renewable Integration: Power Balancing. *IEEE Trans. Power Syst.* **2013**, *28*(4), 4109–4117.

[20] International Energy Agency (IEA). *Batteries and Secure Energy Transitions*; IEA Publications: Paris, 2024.

[21] Zhang, Y.; Lund, P. D.; Wang, J. Challenges and Progresses of Energy Storage Technology and Its Application in Power Systems. *J. Mod. Power Syst. Clean Energy* **2016**, *4* (4), 519–528.

[22] Njema, G. G.; Wang, Z.; Nkosi, V.; Dudelzak, S.; Ozoemena, K. I. A Review on the Recent Advances in Battery Development. *Eng. Rep.* **2024**, *6*, 2329261.

[23] Adeyinka, A. M.; Bamisile, O.; Olagoke, O.; Duvvuru, R.; Olatunji, O. Advancements in Hybrid Energy Storage Systems for Renewable Energy Integration. *Sustain. Energy Res.* **2024**, *9*, 120.

[24] Jose, S. A.; Gallant, A.; Lechuga Gomez, P.; Jaggers, Z.; Johansson, E.; LaPierre, Z.; Menezes, P. L. *Solid-State Lithium Batteries: Advances, Challenges, and Future Perspectives. Batteries* **2025**, *11* (3), 90.

[25] Phogat, P.; Dey, S.; Wan, M. Powering the Sustainable Future: A Review of Emerging Battery Technologies and Their Environmental Impact. *RSC Sustain.* **2025**, *3*, 3266-3306

[26] Conway, B.E. *Electrochemical Supercapacitors: Scientific Fundamentals and Technological Applications*; Springer Science & Business Media: Berlin/Heidelberg, Germany, **2013**.

[27] Miller, J.R.; Simon, P. Electrochemical capacitors for energy management. *Science* **2008**, *321*, 651–652.

[28] Frackowiak, E.; Béguin, F. *Supercapacitors: Materials, Systems and Applications*; Wiley-VCH Verlag GmbH & Co.: Weinheim, Germany, **2013**.

[29] González, A.; Goikolea, E.; Andoni, J.; Mysyk, R. Review on supercapacitors: Technologies and materials. *Renew. Sustain. Energy Rev.* **2016**, *58*, 1189–1206.

[30] Hidalgo-Reyes, J.; Gómez-Aguilar, J.F.; Escobar-Jiménez, R.F.; Alvarado-Martínez, V.M.; López-López, M. Classical and fractional-order modeling of equivalent electrical circuits for supercapacitors and batteries, energy management strategies for hybrid systems, and methods for the state of charge estimation: A state-of-the-art review. *Microelectron. J.* **2019**, *85*, 109–128.

[31] Şahin, M.E.; Blaabjerg, F.; Sangwongwanich, A. A comprehensive review on supercapacitor applications and developments. *Energies* **2022**, *15*, 674.

[32] Queirós, G.; Rey-Raap, N.; Pereira, C.; Pereira, M. CNT-based materials as electrodes for flexible supercapacitors. *U. Porto J. Eng.* **2021**, *7*, 151–162.

[33] Bueno, P.R. Nanoscale origins of super-capacitance phenomena. *J. Power Sources* **2019**, *414*, 420–434.

[34] Tehrani, Z.; Thomas, D.J.; Korochkina, T.; Phillips, C.O.; Lupo, D.; Lehtimäki, S.; O'Mahony, J.; Gethin, D.T. Large-area printed supercapacitor technology for low-cost domestic green energy storage. *Energy* **2017**, *118*, 1313–1321.

[35] Jalal, N.; Ibrahim, R.; Kadhim, M. A review on supercapacitors: Types and components. *J. Phys. Conf. Ser.* **2021**, *1973*, 012015.

[36] Wang, C.; Zhou, E.; He, W.; Deng, X.; Huang, J.; Ding, M.; Wei, X.; Liu, X.; Xu, X. $NiCo_2O_4$-based supercapacitor nanomaterials. *Nanomaterials* **2017**, *7*, 41.

[37] Sun, Z.; Yuan, A. Electrochemical performance of nickel hydroxide/activated carbon super-capacitors using a modified polyvinyl alcohol based alkaline polymer electrolyte. *Chin. J. Chem. Eng.* **2009**, *17*, 150–155.

[38] Sivaraman, P.; Thakur, A.; Kushwaha, R.K.; Ratna, D.; Samui, A.B. Poly(3-methyl thiophene)-activated carbon hybrid supercapacitor based on gel polymer electrolyte. *Electrochem. Solid-State Lett.* **2006**, *9*, A435–A438.

[39] Goodenough, J.B.; Park, K.S. The Li-ion rechargeable battery: A perspective. *J. Am. Chem. Soc.* **2013**, *135*, 1167–1176.

[40] Zhang, H.; Zhang, Y.; Huang, L.; Song, J.; Huang, Z. Study on the influence of connection structure between batteries on battery pack performance. *Electronics* **2024**, *13*, 817.

[41] Whittingham, M.S. Electrical energy-storage and intercalation chemistry. *Science* **1976**, *192*, 1126–1127.

[42] Ghasemi Yeklangi, A.; Ghafari, A.; Asgari Sima, F. Advancing lithium-ion battery manufacturing: Novel technologies and emerging trends. *J. Appl. Electrochem.* **2024**, *54*, 2653–2682.

[43] Schomburg, F.; Heidrich, B.; Wennemar, S.; Drees, R.; Roth, T.; Kurrat, M.; Heimes, H.; Jossen, A.; Winter, M.; Cheong, J.Y.; Roder, F. Lithium-ion battery cell formation: Status and future directions towards a knowledge-based process design. *Energy Environ. Sci.* **2024**, *17*, 2686–2733.

[44] Mizushima, K.; Jones, P.C.; Wiseman, P.C.; Goodenough, J.B. Li_xCoO_2 ($0<x\leq1$): A new cathode material for batteries of high energy density. *Solid State Ion.* **1981**, *3–4*, 171–174.

[45] Thackeray, M.M.; David, W.I.F.; Bruce, P.G.; Goodenough, J.B. Lithium insertion into manganese spinels. *Mater. Res. Bull.* **1983**, *18*, 461–472.

[46] Ebner, W.; Fouchard, D.; Xie, L. The $LiNiO_2$/carbon lithium-ion battery. *Solid State Ion.* **1994**, *69*, 238–256.

[47] Thackeray, M.M.; Johnson, P.J.; De Picciotto, L.A.; Bruce, P.G.; Goodenough, J.B. Electrochemical extraction of lithium from $LiMn_2O_4$. *Mater. Res. Bull.* **1984**, *19*, 179–187.

[48] Feng, L.; Chang, Y.; Wu, L.; Lu, T. Electrochemical behaviour of spinel $LiMn_2O_4$ as positive electrode in rechargeable lithium cells. *J. Power Sources* **1996**, *63*, 149–152.

[49] Murphy, D.W.; Christian, P.A.; Disalvo, F.J.; Carides, J.N. Vanadium oxide cathode materials for secondary lithium cells. *J. Electrochem. Soc.* **1979**, *126*, 497–499.

[50] Lazzari, M.; Scrosati, B. A cyclable lithium organic electrolyte cell based on two intercalation electrodes. *J. Electrochem. Soc.* **1980**, *127*, 773–774.

[51] Ghasemi Yeklangi, A.; Ghafari, A.; Asgari Sima, F. Advancing lithium-ion battery manufacturing: Novel technologies and emerging trends. *J. Appl. Electrochem.* **2024**, *54*, 2653–2682.

[52] Schomburg, F. Lithium-ion battery cell formation: Status and future directions towards a knowledge-based process design. *Energy Environ. Sci.* **2024**, *17*, 2686–2733.

[53] Tarascon, J.M.; Armand, M. Issues and challenges facing rechargeable lithium batteries. *Nature* **2001**, *414*, 359–367.

[54] Xu, K. Electrolytes and interphases in Li-ion batteries and beyond. *Chem. Rev.* **2014**, *114*, 11503–11618.

[55] Armand, M.; Tarascon, J.M. Building better batteries. *Nature* **2008**, *451*, 652–657.

[56] Croce, F.; Appetecchi, G.B.; Persi, L.; Scrosati, B. Nanocomposite polymer electrolytes for lithium batteries. *Nature* **1998**, *394*, 456–458.

[57] Liu, W.; Liu, N.; Sun, J.; Hsu, P.C.; Li, Y.; Lee, H.W.; Cui, Y. Ionic conductivity enhancement of polymer electrolytes with ceramic nanowire fillers. *Nano Lett.* **2015**, *15*, 2740–2745.

[58] Cao, C.; Li, Z.B.; Wang, X.L.; Zhao, X.B.; Han, W.Q. Recent advances in inorganic solid electrolytes for lithium batteries. *Front. Energy Res.* **2014**, *2*, 25.

[59] Quartarone, E.; Mustarelli, P. Electrolytes for solid-state lithium rechargeable batteries: Recent advances and perspectives. *Chem. Soc. Rev.* **2011**, *40*, 2525–2540.

[60] Goodenough, J.B.; Singh, P. Review—Solid electrolytes in rechargeable electrochemical cells. *J. Electrochem. Soc.* **2015**, *162*, A2387–A2392.

[61] Stephan, A.M.; Nahm, K.S. Review on composite polymer electrolytes for lithium batteries. *Polymer* **2006**, *47*, 5952–5964.

[62] Kulkarni, A.R. Solid state ionics, Proc. Asian Conf., 8th. *Solid State Ionics* **2002**, 273–282.

[63] Zhang, H.; Wang, P.; Li, Y.; Fu, J. Strategies toward scalable composite solid electrolytes for practical solid-state lithium batteries. *Nano Energy* **2022**, *99*, 107419.

[64] Yang, Y.; Xie, D.; Yang, Q.; Li, Y.; Liu, W. Composite solid electrolytes: Materials design and perspectives for all-solid-state lithium batteries. *Adv. Funct. Mater.* **2023**, *33*, 2209745.

[65] Xia, Y.; Zheng, J.; Goodenough, J.B. Review on composite solid-state electrolytes for lithium batteries. *Chem. Rev.* **2021**, *121*, 1331–1392.

[66] Zhu, M.; Wu, J.; Wang, Y.; Song, M.; Long, L.; Siyal, S.H.; Yang, X.; Sui, G. Recent advances in gel polymer electrolyte for high-performance lithium batteries. *J. Energy Chem.* **2019**, *37*, 126–142.

[67] Ngai, K.S.; Ramesh, S.; Ramesh, K. A review of polymer electrolytes: Fundamental, approaches and applications. *Ionics* **2016**, *22*, 1259–1279.

[68] Mahendran, M.; Mohamed, R.S.; Sundaram, N.S.; Ahmad, A. Enhanced ionic conductivity and stability of plasticized PVC-based polymer electrolytes for electrochemical devices. *J. Polym. Res.* **2023**, *30*, 255.

[69] Zhao, Q.; He, X.; Liu, Y.; Chen, R. Enhancing ion transport in plasticized polymer electrolytes using eutectic plasticizer systems. *Electrochim. Acta* **2022**, *423*, 140570.

[70] Kirana, M.G.; Jyothib, N.K.; Rao, M.C. Structural, electrical and dielectric studies of PVA-based $NaNO_3$ polymer electrolytes for battery applications. *Int. J. Ambient Energy* **2023**, *44*, 246–255.

[71] Malathi, J.; Kumaravadivel, M.; Brahmanandhan, G.M. Structural, thermal and electrical properties of PVA–$LiCF_3SO_3$ polymer electrolyte. *J. Non-Cryst. Solids* **2010**, *356*, 2277–2281.

[72] Roy, A.S.; Gupta, S.; Sindhu, S.; Parveen, A.; Ramamurthy, P.C. Dielectric properties of novel PVA/ZnO hybrid nanocomposite films. *Compos. Part B* **2013**, *47*, 314–319.

[73] Radha, K.P.; Selvasekarapandian, S.; Karthikeyan, S.; Hema, M.; Sanjeeviraja, C. Synthesis and impedance analysis of proton-conducting polymer electrolyte PVA:NH_4F. *Ionics* **2013**, *19*, 1437–1447.

[74] Hema, M.; Selvasekarapandian, S.; Arunkumar, D.; Sakunthala, A.; Nithya, H. FTIR, XRD and ac impedance spectroscopic study on PVA-based polymer electrolyte doped with NH_4X (X = Cl, Br, I). *J. Non-Cryst. Solids* **2009**, *355*, 84–90.

[75] Yusof, Y.M.; Illias, H.A.; Kadir, M.F.Z. Incorporation of NH_4Br in PVA-chitosan blend-based polymer electrolyte and its effect on the conductivity and other electrical properties. *Ionics* **2014**, *20*, 1235–1245.

[76] Agarwal, S.L.; Awadhia, A. DSC and conductivity studies on PVA-based proton conducting gel electrolytes. *Bull. Mater. Sci.* **2004**, *27*, 523–527.

[77] Yang, C.-C. Synthesis and characterization of the cross-linked PVA/TiO_2 composite polymer membrane for alkaline DMFC. *J. Membr. Sci.* **2007**, *288*, 51–60.

[78] Mohan, V.M.; Qiu, W.; Shen, J.; Chen, W. Electrical properties of poly(vinyl alcohol) (PVA) based $LiFePO_4$ complex polymer electrolyte films. *J. Polym. Res.* **2010**, *17*, 143–150.

[79] Vijaya, N.; Selvasekarapandian, S.; Karthikeyan, S.; Prabu, M.; Rajeswari, N.; Sanjeeviraja, C. Synthesis and characterization of proton conducting polymer electrolyte based on poly(N-vinyl pyrrolidone). *J. Appl. Polym. Sci.* **2013**, *127*, 1538–1543.

[80] Ravi, M.; Pavani, Y.; Kiran Kumar, K.; Bhavani, S.; Sharma, A.K.; Narasimha Rao, V.V.R. Studies on electrical and dielectric properties of PVP:$KBrO_4$ complexed polymer electrolyte films. *Mater. Chem. Phys.* **2011**, *130*, 442–448.

[81] Ravi, M.; Bhavani, S.; Pavani, Y.; Narasimha Rao, V.V.R. Investigation on electrical properties of PVP:$KClO_4$ polymer electrolyte films. *Ind. J. Pure Appl. Phys.* **2013**, *51*, 362–366.

[82] Ramya, C.S.; Selvasekarapandian, S.; Savitha, T.; Hirankumar, G.; Baskaran, R.; Bhuvaneswari, M.S.; Angelo, P.C. Conductivity and thermal behavior of proton conducting polymer electrolyte based on poly(N-vinyl pyrrolidone). *Eur. Polym. J.* **2006**, *42*, 2672–2677.

[83] Vijaya, N.; Selvasekarapandian, S.; Hirankumar, G.; Karthikeyan, S.; Nithya, H.; Ramya, C.S.; Prabu, M. Structural, vibrational, thermal, and conductivity studies on proton-conducting polymer electrolyte based on poly(N-vinylpyrrolidone). *Ionics* **2012**, *18*, 91–99.

[84] Shahenoor Basha, S.K.; Sunita Sundari, G.; Vijay Kumar, K.; Rao, M.C. Structural and dielectric properties of PVP-based composite polymer electrolyte thin films. *J. Inorg. Organomet. Polym.* **2017**, *27*, 455–466.

[85] Ravi, M.; Kiran Kumar, K.; Madhu Mohan, V.; Narasimha Rao, V.V.R. Effect of nano TiO_2 filler on the structural and electrical properties of PVP-based polymer electrolyte films. *Polym. Test.* **2014**, *33*, 152–160.

[86] Shahenoor Basha, S.K.; Sunita Sundari, G.; Vijay Kumar, K.; Veera Bhadra Reddy, K.; Rao, M.C. Electrical conduction behaviour of PVP-based composite polymer electrolytes. *Rasayan J. Chem.* **2017**, *10*, 279–285.

[87] Sengwa, R.J.; Choudhary, S.; Sankhla, S. Low frequency dielectric relaxation processes and ionic conductivity of montmorillonite clay nanoparticles colloidal suspension in poly(vinyl pyrrolidone)-ethylene glycol blends. *Express Polym. Lett.* **2008**, *2*, 800–809.

[88] Daries Bella, R.S.; Hirankumar, G.; Devaraj, P. Characterization of plasticized proton conducting polymer electrolyte and its application in primary proton battery. *Int. J. Chem. Res.* **2014**, *6*, 5372–5377.

[89] Modified thermal, dielectric, and electrical conductivity of PVDF-HFP/$LiClO_4$ polymer electrolyte films by 8 MeV electron beam irradiation. *ACS Omega* **2018**, *3*, 14188–14200.

[90] Liu, W. A study on PVDF-HFP gel polymer electrolyte for lithium-ion batteries. *IOP Conf. Ser.: Mater. Sci. Eng.* **2017**, *213*, 012036.

[91] Ahmad, A.L.; Farooqui, U.R.; Hamid, N.A. Porous (PVDF-HFP/PANI/GO) ternary hybrid polymer electrolyte membranes for lithium-ion batteries. *RSC Adv.* **2018**, *8*, 25725.

[92] Du, S.-Y.; Ren, G.-X.; Zhang, N.; Liu, X.-S. High-performance PVDF-HFP based composite electrolytes with excellent interfacial compatibility for room-temperature all-solid-state lithium metal batteries. *ACS Omega* **2022**, *7*, 19631–19639.

[93] Halima, N.B. Poly(vinyl alcohol): Review of its promising applications and insights into biodegradation. *RSC Adv.* **2016**, *6*, 39811–39832.

[94] Khan, M.M.R.; Rumon, M.M.H.; Islam, M.S. Synthesis, rheology, morphology, and mechanical properties of biodegradable PVA-based composite films: A review on recent progress. *Processes* **2024**, *12*, 2880.

[95] Ha, S.-H.; Chang, Y.-C.; Lai, J.-Y.; Hou, S.; Lee, K.-R. Alkaline solid polymer electrolyte membranes based on structurally modified PVA/PVP. *Polymer* **2010**, *51*, 5360–5369.

[96] Deshmukh, K.; Ahamed, M.B.; Polu, A.R.; Sadasivuni, K.K.; Pasha, S.K.K.; Ponnamma, D.; Al-Ali AlMaadeed, M.; Deshmukh, R.R.; Chidambaram, K. Impedance spectroscopy, ionic conductivity and dielectric studies of new Li^+ ion conducting polymer blend electrolytes based on biodegradable polymers for solid-state battery applications. *J. Mater. Sci.: Mater. Electron.* **2016**, *27*, 11410–11424.

[97] Rajeswari, N.; Selvasekarapandian, S.; Prabu, M.; Karthikeyan, S.; Sanjeeviraja, C. Lithium-ion conducting solid polymer blend electrolyte based on biodegradable polymers. *Bull. Mater. Sci.* **2013**, *36*, 333–339.

[98] Hatta, F.F.; Yahya, M.Z.A.; Ali, A.M.M.; Subban, R.H.Y.; Harun, M.K.; Mohamad, A.A. Electrical conductivity studies on PVA/PVP-KOH alkaline solid polymer blend electrolyte. *Ionics* **2005**, *11*, 418–422.

[99] Rajeswari, N.; Selvasekarapandian, S.; Karthikeyan, S.; Sanjeeviraja, C.; Iwai, Y.; Kawamura, J. Structural, vibrational, thermal, and electrical properties of PVA/PVP biodegradable polymer blend electrolyte with CH_3COONH_4. *Ionics* **2013**, *19*, 1105–1113.

[100] Rajeswari, N.; Selvasekarapandian, S.; Karthikeyan, S.; Nithya, H.; Sanjeeviraja, C. Lithium-ion conducting polymer electrolyte based on poly(vinyl alcohol)–poly(vinyl pyrrolidone) blend with $LiClO_4$. *Int. J. Polymeric Mater. Polym. Biomater.* **2012**, *61*, 1164–1175.

[101] Karpagavel, K.; Sundaramahalingam, K.; Manikandan, A.; Vanitha, D.; Manohar, A.; Nagarajan, E.R.; Nallamuthu, N. Electrical properties of lithium-ion conducting PVDF-HFP/PVP solid polymer electrolyte. *J. Electron. Mater.* **2021**, *50*, 4415–4425.

[102] Wang, Z.; Tang, Z. Characterization of blend PVDF-HFP/PVP polymer electrolyte with $LiBF_4$. *Solid State Ion.* **2003**, *82*, 273–282.

[103] Saikia, D.; Kumar, A. Ionic Conduction in P(VDF-HFP)/PVDF–(PC + DEC)–$LiClO_4$ Polymer Gel Electrolytes. *Electrochim. Acta* **2004,** *49*, 2581–2589.

Chapter 2

Experimental Techniques

2.1 Introduction

Solid polymer electrolytes (SPEs) represent a promising category of materials offering significant potential for application in diverse electrochemical systems, particularly solid-state batteries to their safety, flexibility, and tuneable conductivity. The preparation techniques of a SPE play a crucial role in dictating its morphology, degree of crystallinity, salt dispersion, and ultimately, ionic conductivity and electrochemical stability [1]. Numerous processing approaches have been designed, each tailored to attain the desired characteristics of the host polymer and application demands.

2.1.1 Solution Casting Technique

Among the various methods, solution casting is a commonly employed and simple approach for fabricating SPE films, valued for its ease and adaptability [2,3]. This technique entails dissolving a polymer along with an ionic salt in an appropriate solvent, then casting the solution and allowing gradual solvent evaporation to obtain thin films.

Solvent Selection Criticality: The choice of solvent is very important [4]. It must undergo effective dissolution of all components, polymers, and salts without inducing chemical degradation and must have a suitable boiling point for efficient evaporation. Commonly used solvents include water, acetonitrile, tetrahydrofuran (THF), and N, N-dimethylformamide (DMF), chosen based on their volatility and polarity. The solvent's polarity and its interaction with the polymer chains influence the final morphology and crystallinity [5].

Homogeneous Mixing: Achieving a truly homogeneous solution is critical. A homogeneous SPE matrix can be achieved through continuous stirring and sometimes with gentle heating, ensuring the polymer chains are fully solvated and the salt is completely dissociated and uniformly dispersed. Incomplete dissolution can lead to localized regions of high crystallinity or undissolved salt, creating pathways for non-uniform ionic transport and reduced performance.

Controlled Evaporation: The rate of solvent evaporation significantly impacts the final film's morphology, porosity, and mechanical strength [6]. Slow Evaporation is generally preferred, as it allows polymer chains to rearrange and pack more efficiently, leading to denser, less porous films with potentially higher mechanical integrity and better salt dispersion. Fast

Evaporation can lead to more porous films, which might increase ionic pathways but often at the expense of mechanical strength and can trap residual solvent.

Substrate Choice: The substrate on which the solution is cast (e.g., Teflon, glass, polypropylene) influences the film's release and surface roughness [7]. Non-stick surfaces are preferred for easy detachment.

Post-Drying/Annealing: After initial drying, further vacuum drying at elevated temperatures is often employed to remove residual solvent completely, as even trace amounts can plasticize the polymer and degrade performance over time. Heating above the glass transition temperature (Tg) of the polymer enhances chain flexibility, which facilitates improved salt distribution and increases crystallinity, ultimately boosting ionic conductivity.

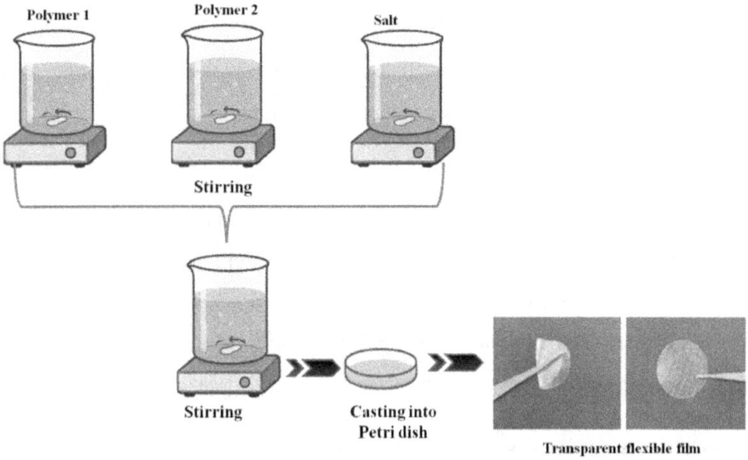

Figure 2.1 Solution casting method for preparation of solid polymer electrolytes.

2.1.2 Hot Pressing (Melt Pressing) Technique

Hot pressing is a solvent-free technique, making it environmentally friendly and desirable for large-scale production [8]. It relies on the thermoplastic nature of the polymer.

Polymer Melt Viscosity: This technique relies strongly on the melt viscosity of the polymer, which needs to attain adequate fluidity at the pressing temperature to uniformly embed the salt. Polymers with high melt viscosity might require higher temperatures or pressures, which could lead to thermal degradation of the polymer or the salt.

Temperature and Pressure Control: Precise control over both temperature and pressure is crucial.

> **Temperature:** Must be above the polymer's melting point (Tm) (for crystalline polymers) or glass transition temperature (Tg) (for amorphous polymers) to allow for sufficient flow. Too high a temperature can cause degradation; too low, insufficient flow.
>
> **Pressure:** Ensures good contact between polymer and salt particles and promotes densification of the film, minimizing voids. The applied pressure helps to homogenize the mixture and eliminate air bubbles.

Homogenization of Components: Before pressing, the polymer and salt (and any other additives) are typically mixed thoroughly in powder form. Techniques like ball milling or twin-screw extrusion are often used to achieve a fine, uniform mixture, preventing localized salt aggregation which can lead to poor ionic pathways [9].

Film Thickness Control: The final film thickness is governed by the quantity of material employed and the pressure exerted during pressing, allowing the fabrication of films with precise dimensions.

Absence of Solvent-Related Issues: Eliminates concerns about residual solvent, solvent-polymer interactions, and the environmental impact of solvent disposal. However, it can introduce issues related to thermal degradation if not carefully controlled.

2.1.3 Electro spinning

Electrospinning is a flexible method for generating continuous nanofibers, providing a large surface-area-to-volume ratio that benefits ion transport [10].

Solution Rheology: Stable electrospinning strongly depends on the rheological characteristics of the polymer solution, such as viscosity and surface tension [11]. The solution must be viscous enough to form continuous fibers but not so viscous that it clogs the spinneret. Adjusting polymer concentration, molecular weight, and solvent composition is essential.

High Voltage and Electric Field: The technique utilizes a high electric field to elongate and pull the polymer solution into thin fibers [12]. Fiber diameter and morphology are strongly affected by parameters such as applied voltage, the gap between the spinneret and collector, and the solution flow rate [13]. Too high a voltage can lead to "beads-on-string" morphology, while too low may not initiate fiber formation.

Nanofiber Morphology: Electrospun SPEs can have unique advantages due to their fibrous structure:

> **Increased Interfacial Area:** The high surface area of nanofibers can potentially enhance electrode-electrolyte contact in solid-state devices.
>
> **Open Porous Structure:** The interconnected network of fibers can create pathways for ion migration, potentially improving ionic conductivity, especially at lower polymer crystallinity [14].
>
> **Anisotropic Properties:** Fibers might exhibit anisotropic ionic conductivity depending on their alignment [15].
>
> **Scalability Challenges:** While promising, scaling up electrospinning for large-scale production of dense, pinhole-free SPE membranes can be challenging compared to solution casting or hot pressing.

2.1.4 In-situ Polymerization

In-situ polymerization involves polymerizing the monomer directly in the presence of the salt and other components, often within the electrochemical cell itself [16].

Conformational Adaptability: As the polymer forms around the salt, it can potentially create a more intimate and stable interface, leading to better salt dissolution and more efficient ion transport pathways. The growing polymer chains can adapt to the presence of the salt.

Reduced Interfacial Resistance: This approach is intended to lower the interfacial resistance between electrodes and the electrolyte by forming the electrolyte in situ on the electrode surface.

Monomer Selection: The choice of monomer is crucial. It must polymerize efficiently under specific conditions (e.g., UV, thermal initiation) and form a polymer with desirable properties for ion conduction. Common monomers include acrylates, methacrylates, and epoxides.

Initiator and Curing Conditions: The type and concentration of initiator and the curing conditions directly influence the polymerization rate, degree of crosslinking, and final mechanical properties.

Volume Shrinkage: Polymerization often involves volume shrinkage, which can lead to stress within the electrolyte and at the electrode interface. Careful formulation can minimize this.

2.1.5 Electrodeposition Polymerization:

This method involves the electrochemical polymerization of monomers directly onto an electrode surface.

Direct Film Formation on Electrode: The primary advantage is the direct formation of a uniform, thin film of SPE on the electrode surface, potentially leading to excellent interfacial contact and reduced resistance.

Control over Film Thickness: The film thickness can be accurately regulated by varying the electrode position duration, applied current density, and the concentration of monomers.

Mechanism of Polymerization: This typically involves oxidative polymerization of monomers (e.g., pyrrole, aniline, thiophene derivatives, or even some anilines with pendant ether chains for ion conduction). The resulting polymers are often intrinsically conductive, which might be a disadvantage for purely ion-conducting electrolytes, but can be tailored for hybrid systems.

Pore Structure and Morphology: The morphology of the deposited film can be influenced by the electrochemical conditions, potentially leading to porous structures that enhance ion transport.

Limited Monomer Scope: The range of monomers that can be electrochemically polymerized into ionically conductive SPEs is more limited compared to other techniques.

Scalability Challenges: While promising for small-scale applications, scaling up electrodeposition for large-area electrolytes can be challenging.

Every preparation method presents its own benefits and limitations, with the selection largely determined by the polymer–salt system, the required film characteristics (such as thickness, mechanical stability, and conductivity), and the scalability needed for the intended application.

In this study, solid polymer electrolyte (SPE) films were fabricated through the solution casting method, which enables homogeneous salt distribution in the polymer matrix, accurate regulation of film thickness, and convenient fabrication of defect-free, flexible membranes ideal for electrochemical evaluation.

2.2 Polymer electrolyte characterization techniques

2.2.1 X-ray Diffraction (XRD)

X-ray Diffraction (XRD) is a robust, non-destructive method extensively employed to analyze the crystalline structure of materials, particularly SPEs.

2.2.1.1 Introduction to XRD

XRD functions based on **Bragg's Law**, which explains that constructive interference arises at certain angles when X-rays strike the periodically arranged atomic planes of a crystal. This phenomenon, known as diffraction, produces a unique "fingerprint" pattern of diffracted X-rays that is characteristic of the material's crystal structure. Analyzing the angles and intensities of the diffracted X-rays provides researchers with important information about the crystallographic characteristics of the material. The scattered radiation reveals detailed information about the arrangement of the atoms in the material. Figure 2.2 shows the visual representation of Bragg's law.

A standard PXRD instrument comprises:

X-ray Source: An X-ray tube (e.g., copper target) generates monochromatic X-rays.

X-ray Optics: Slits and sometimes mirrors condition the X-ray beam.

Goniometer: A precise mechanical system (often Bragg-Brentano geometry) that controls the angular positions of the X-ray source, sample, and detector to scan through 2θ angles.

Sample Holder: Securely positions the sample (film or powder) for accurate analysis.

X-ray Detector: Measures the intensity of diffracted X-rays (e.g., point, linear, or area detectors).

Data Acquisition and Analysis System: Controls the instrument, collects data (intensity vs. 2θ), and processes it for phase identification, crystallinity quantification, and structural refinement.

2.2.1.2 Working Principle

The principle behind XRD lies in the interaction of X-rays with the crystal lattice, where most X-rays pass through but some are scattered, forming a distinctive interference pattern. The basic concept underlying X-ray diffraction (XRD) is Bragg's law, which can be mathematically represented as:

$$n\lambda = 2d\sin\theta \quad \text{.......... 2.1}$$

where:

n is the diffraction order (an integer),

λ is the wavelength of the X-rays,

d is the spacing between atomic planes in the crystal,

θ is half the scattering angle 2θ.

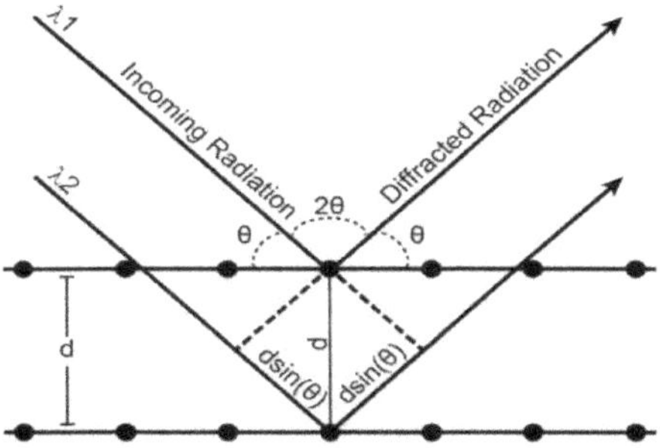

Figure 2.2 Visual representation of Bragg's law [17].

When a monochromatic X-ray beam (with wavelength λ) strikes a crystalline sample, it interacts with the electrons of the atoms within the crystal lattice. If the X-rays hit atomic planes that are spaced at a distance 'd' apart and the angle of incidence (θ) satisfies Bragg's Law, constructive interference occurs. This constructive interference results in a diffracted X-ray beam of high intensity as shown in Figure 2.2.

In a powdered sample, the random arrangement of many crystallites ensures that enough are aligned at the appropriate θ angle to fulfil Bragg's Law for every set of atomic planes. When the sample is rotated, or the X-ray source and detector are swept across a range of angles (commonly 2θ), the detector measures the intensity of the diffracted X-rays. The resulting plot of intensity versus 2θ (the diffractogram) shows sharp peaks at specific 2θ values corresponding to the unique interplanar spacing's (d) within the crystalline phases of the material. Amorphous materials, lacking long-range atomic order, produce broad, diffuse patterns (often referred to as an "amorphous halo") rather than sharp peaks [18].

2.2.1.3 Powder X-ray diffraction (PXRD)

Powder X-ray diffraction is the most frequently used XRD technique and is ideal for polycrystalline or powdered samples. Since the sample contains many randomly oriented crystallites, X-rays are diffracted in various directions, producing a pattern that can be analysed for phase identification, quantification, and lattice parameter determination.

PXRD is particularly invaluable for characterizing solid polymer electrolytes (SPEs) due to their often-semi-crystalline nature. Unlike single-crystal XRD, which examines a single, perfectly ordered crystal, powder XRD analyses a sample composed of a vast number of tiny, randomly oriented crystallites. This allows for the study of bulk materials, making it highly practical for the diverse forms in which SPEs are prepared. Each crystalline substance yields a distinct diffraction pattern, characterized by the positions and intensities of the peaks, this makes PXRD a critical tool in evaluating the degree of crystallinity, salt–polymer interactions, and the impact of additives or processing conditions on the structural orientation of SPEs.

2.2.1.4 Sample Preparation of SPEs for PXRD analysis

Sample preparation for XRD of SPEs is generally straightforward, but critical for obtaining good quality data:

a) Film Samples: The SPE films prepared via solution casting techniques are typically cut into small, flat pieces (often a few square centimetres) and mounted onto a zero-background sample holder (often made of silicon or glass with a recessed well) [19]. It's essential to ensure the film lies perfectly flat to prevent peak shifts or broadening caused by an uneven surface. For very thin films, stacking multiple layers can enhance the signal.

b) Powder Samples: If the SPE is brittle or prepared in powdered form (e.g., after grinding), it can be finely ground to a uniform powder. This powder is then typically packed into a shallow well on a sample holder or dispersed on a non-diffracting substrate. Ensuring a random orientation of crystallites in a powder is important for accurate intensity measurements.

c) Removal of Residual Solvent: For SPEs prepared via solution-cast films, complete removal of residual solvent is essential [20], as solvent molecules can interfere with the polymer's crystalline packing and lead to erroneous XRD results. Vacuum drying is often employed.

d) Contamination Avoidance: All sample preparation steps should minimize contamination, as even small amounts of crystalline impurities can introduce extraneous peaks in the XRD pattern.

2.2.1.5 Role of XRD in Characterization of Solid Polymer Electrolytes

XRD is essential for analyzing SPEs, as it reveals their structural organization, which strongly influences ionic conductivity and overall performance.

a) Crystallinity Determination: Polymers can be amorphous, crystalline, or semi-crystalline. In SPEs, ionic transport mainly takes place within the amorphous domains, where the flexibility of polymer chains facilitates ion migration through hopping [21]. XRD is used to quantify the degree of crystallinity in SPEs [22]. A highly crystalline polymer host will exhibit sharp, intense peaks, while amorphous regions will contribute to a broad hump. By analyzing the relative areas of crystalline peaks and the amorphous halo, the percentage of crystallinity can be estimated. A reduction in crystallinity upon salt addition or incorporation of plasticizers/fillers (often observed as a decrease in peak intensity and an increase in the amorphous halo) is a positive indication for enhanced ionic conductivity [23].

b) Salt Dissolution and Complexation: XRD can confirm the complete dissolution of the lithium or other ionic salts within the polymer matrix. If the salt remains undissolved, its characteristic sharp diffraction peaks will be present in the XRD pattern. The disappearance or significant reduction of these salt peaks indicates successful dissolution and complexation with the polymer, which is vital for free ion movement.

In summary, X-ray diffraction (XRD) serves as an essential technique for analyzing the key structural features of solid polymer electrolytes, offering valuable understanding of how composition and fabrication influence their degree of crystallinity, salt incorporation, and, consequently, the mechanisms of ionic transport.

The XRD analysis in this study was performed using a Bruker X-ray diffractometer equipped with CuKα radiation (λ = 1.540 Å), operating at a scan rate of 5°/min within the range of 10°–80°. The XRD setup is illustrated in Figure 2.3.

Figure 2.3 XRD instrument photograph.

2.2.2 Fourier Transform Infrared Spectroscopy (FTIR) Studies

FTIR (Fourier Transform Infrared) spectroscopy is a powerful analytical method used to investigate the distinct vibrational "fingerprints" of molecules [24]. Unlike other spectroscopic methods that use a single wavelength, FTIR bombards the sample with a broad spectrum of infrared light simultaneously, capturing information across a wide range of frequencies in a swift and highly sensitive manner [25]. This makes it an indispensable tool for identifying functional groups, understanding molecular interactions, and probing the chemical changes within materials, including the intricate world of SPEs.

2.2.2.1 Working Principle: Molecular Vibrations and the Interferogram

The fundamental concept of FTIR spectroscopy is based on the interaction between infrared (IR) radiation and molecular vibrations. When IR light passes through a sample, certain frequencies are absorbed by the molecules, causing their specific chemical bonds to vibrate (stretch, bend, wag, twist). A vibration becomes IR active only when it induces a change in the dipole moment of the molecule. The specific frequencies absorbed are unique to the functional groups within the molecule, serving as its distinctive "molecular fingerprint."

Unlike older dispersive IR spectrometers that scan through frequencies sequentially, FTIR employs a Michelson interferometer [26]. This sophisticated instrument divides a single IR beam into two pathways: one directed toward a stationary mirror and the other toward a movable mirror [26]. When the two beams merge again, they generate an interference pattern known as an interferogram [26]. This interferogram is a complex signal representing the sum of all IR frequencies simultaneously. A mathematical operation called a Fourier Transform then converts this time-domain interferogram into the familiar frequency-domain spectrum (intensity vs. wavenumber, cm^{-1}), revealing the distinct absorption bands of the sample. This "multiplex advantage" allows for faster data acquisition and higher signal-to-noise ratios compared to dispersive instruments.

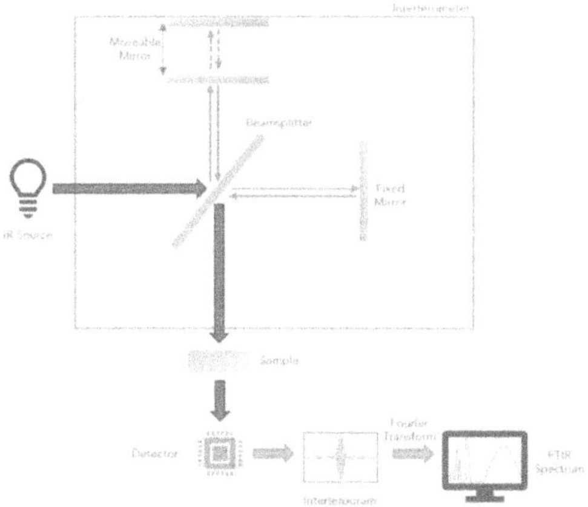

Figure 2.4 Schematic diagram of a Michelson interferometer configured for FTIR.

A typical FTIR spectrometer comprises several key components that work together:

a) IR Source: A broadband source, such as a heated silicon carbide (Globar) rod or a nichrome wire, emits infrared radiation across a wide range of wavelengths.

b) Michelson Interferometer: This is the heart of the FTIR. It consists of:

Beam splitter: A semi-transparent material (often KBr or CaF_2) that splits the incoming IR beam into two paths.

Fixed Mirror: Reflects one beam back to the beam splitter.

Moving Mirror: Reflects the other beam back to the beam splitter. The continuous movement of this mirror creates the path difference between the two beams, generating the interferogram.

c) Sample Compartment: Where the sample is placed. This is highly versatile, accommodating various sampling accessories.

d) Detector: Converts the IR signal into an electrical signal. Common detectors include Deuterated Triglycine Sulphate (DTGS) for routine applications and Mercury Cadmium Telluride (MCT) for higher sensitivity and speed, especially for dilute samples or rapid scans.

e) Computer and Software: Fourier Transform Infrared (FTIR) spectrometers utilize advanced microprocessors with high-speed data acquisition and substantial memory to collect multiple scans and convert the resulting interferogram into an absorbance spectrum plotted against wavenumber. Unlike traditional dispersive spectrometers, which operate in a double-beam configuration to automatically correct for background signals, FTIR systems typically use a single-beam setup. In this approach, two spectra are acquired-one for the background and one for the sample-and the background is digitally subtracted to isolate the sample's spectral features.

2.2.2.2 Sample Preparation for FTIR analysis:

The beauty of FTIR lies in its versatility, offering numerous sample preparation techniques tailored to the specific form of the SPE:

a) Thin Film (Transmission Mode): This is ideal for SPEs prepared as transparent or semi-transparent films (e.g., via solution casting or hot pressing). A small piece of the film is simply placed directly in the IR beam path. For very thick or opaque films, the film might need to be thinned or a different sampling technique employed. Ensuring the film is uniform in thickness is important for quantitative analysis.

b) KBr Pellet (Transmission Mode): For powdered SPE samples or those that are too opaque for direct transmission, the sample is finely ground and mixed with optical grade potassium bromide (KBr) powder. This mixture is then pressed under high pressure into a transparent pellet. KBr is chosen because it is transparent to IR radiation across the typical mid-IR range.

2.2.2.3 Role of FTIR in Characterization of Solid Polymer Electrolytes

FTIR spectroscopy provides a reliable approach to examine the intricate chemistry and interactions within SPEs, yielding essential information about their overall performance:

a) Identification of Polymer-Salt Complex: This is perhaps the most critical application. When a salt (e.g., lithium salt) is incorporated into a polymer matrix, the ions interact with the functional groups of the polymer (e.g., ether oxygens in PEO, carbonyl groups in PMMA). Such interactions lead to variations in the vibrational frequencies of the polymer's distinctive bands. For example, in poly(ethylene oxide) (PEO), the C–O–C stretching vibrations shift to lower wavenumbers when complexed with Li^+ ions, signifying the formation of coordination bonds. Similarly, shifts in carbonyl (C=O) or amine (N–H) bands in other polymers signal interaction. FTIR can therefore confirm the successful dissolution of the salt and the formation of a polymer-ion complex, which is essential for ionic conductivity.

b) Probing Interfacial Interactions in Composites: In composite polymer electrolytes with ceramic fillers (such as SiO_2 or Al_2O_3), FTIR is capable of identifying interactions between the polymer matrix and the filler surface. New bands or shifts in existing bands can indicate hydrogen bonding, Lewis acid-base interactions, or other chemical bonding between the polymer and the filler, influencing chain mobility and ion transport.

c) Studying Polymer Chain Conformation and Crystallinity Changes: While not as direct as XRD for crystallinity, FTIR can indirectly provide information about changes in polymer chain conformation. For example, specific bands in PEO are sensitive to its helical crystalline conformation. A reduction in the intensity of these "crystalline" bands, along with an enhancement of the "amorphous" bands upon salt incorporation or heating, can indicate decreased crystallinity [27].

In essence, FTIR spectroscopy serves as a molecular detective, providing unparalleled insights into the specific chemical bonds and interactions that govern the behaviour of solid polymer electrolytes. Its ability to reveal these microscopic details is fundamental to designing and optimizing SPEs for advanced energy storage applications.

In this study, FTIR transmittance spectra of the polymer electrolyte films were obtained using a **Shimadzu IRTracer-100** spectrometer. The measurements were conducted at a spectral range of 400–4000 cm^{-1}, enabling precise identification of functional groups and polymer–salt interactions.

2.2.3 AC impedance analysis

Electrochemical Impedance Spectroscopy (EIS), often referred to as AC Impedance Spectroscopy, is a sophisticated and adaptable technique used to investigate the complex electrical properties of electrochemical systems, playing a crucial role in the study of solid polymer electrolytes (SPEs) for all-solid-state battery applications.

The method involves applying a small sinusoidal voltage-typically in the range of 5-10 mV peak-to-peak-across the electrodes of an electrochemical cell and measuring the resulting current response. This interaction adheres to Ohm's law in the time domain, Impedance is described as the quotient of voltage divided by current

$$Z = V/I \qquad ....2.2$$

However, unlike simple resistance, impedance is a frequency-dependent complex quantity that encapsulates both resistive (real part, Z') and reactive (imaginary part, Z" due to capacitance or inductance) components. This dual nature allows EIS to capture both energy dissipation and storage mechanisms within the material [28]. By sweeping the AC signal across a wide range of frequencies, EIS constructs a detailed impedance spectrum that reflects the dynamic processes occurring within the electrolyte and at its interfaces [29].Such measurements enable researchers to represent the system with equivalent electrical circuits consisting of resistors (R), capacitors (C), and inductors (L), commonly illustrated using Nyquist plots (imaginary impedance, Z" vs. real impedance, Z') or Bode plots (impedance magnitude and phase angle vs. frequency). Such modelling yields critical parameters including charge transfer resistance, double-layer capacitance, and ionic conductivity, which are essential for understanding the electrochemical kinetics and stability of SPEs.

One of the key advantages of EIS over direct current (DC) techniques is unlike DC methods that can induce significant polarization and distort the true electrochemical response; the alternating current (AC) nature of EIS minimizes such artefacts. The phase difference between voltage and current in AC signals helps isolate individual processes, offering a cleaner and more accurate depiction of the system's intrinsic properties.

Furthermore, EIS can be performed over a range of temperatures, enabling the study of thermally activated conduction mechanisms, ion mobility, and interfacial resistance under varying thermal conditions. This temperature-dependent impedance profiling provides insight into activation energies and the stability of the electrolyte under operational stress [30].

Additionally, EIS facilitates the frequency-resolved assessment of ionic migration through the bulk polymer matrix, charge exchange at electrode–electrolyte interfaces, and transport across grain boundaries in heterogeneous solid-state systems. Through the examination of the in-phase (real part, Z') and out-of-phase (imaginary part, Z") components of the current response, EIS provides a detailed insight into the electrochemical characteristics of SPEs. The impedance (Z) or admittance(Y) of complex amounts can be created as:

$$Z^* = Z'(\omega) - j\, Z''(\omega) \quad \text{......2.3}$$

$$Y^* = G(\omega) + j\, B(\omega) \quad \text{...... 2.4}$$

The normalized terms Z'/G(ω) and Z"/B(ω) represent the real and imaginary components of impedance and admittance, respectively. When represented in the complex plane, these parameters produce frequency dispersion plots that reflect various electrochemical phenomena. Such curves provide valuable insight into the nature of the electrode–electrolyte interactions, intrinsic material resistance, and grain boundary effects [31].

Moreover, additional frequency-dependent parameters—such as complex dielectric constant (ε) and electrical modulus (M)—can be expressed by

$$\varepsilon^*(\omega) = [j\, \omega C_o Z^*(\omega)]^{-1} \quad \text{...... 2.5}$$

$$M^*(\omega) = \varepsilon^*(\omega)^{-1} \quad \text{...... 2.6}$$

where, ω= angular frequency, Co= Vacuum capacitance of the measuring sample

$$C_o = \varepsilon_o (A/d) \quad \text{....... 2.7}$$

where, εo is the permittivity of free space (8.854×10^{-12} F/m)

2.2.3.1 Role of EIS in Characterization of Solid Polymer Electrolytes

EIS is a cornerstone technique for SPE characterization because it provides quantitative and mechanistic insights into their performance. The dynamic characteristics of an electrochemical system can be characterized by analyzing impedance spectra obtained across a broad frequency domain. This process typically involves applying an alternating voltage or current signal and monitoring the system's electrical response.

A complete assessment of the system's electrochemical behaviour requires evaluating both the real (in-phase) and imaginary (out-of-phase) components of impedance across different frequencies within the investigated range. The impedance obtained, represented as a vector, is characterized by the ratio of the voltage phasor to the current phasor.

$$Z_{tot} = (E' + j E'')/(I' + j I'') \qquad \text{....... 2.8}$$

$$Z_{tot} = (Z' + j Z'') \qquad \text{....... 2.9}$$

$$E_{tot} = (E' + j E'') \qquad \text{....... 2.10}$$

$$I_{tot} = (I' + j I'') \qquad \text{....... 2.11}$$

The detectable voltage and current components—namely the real (E′, I′) and imaginary (E″, I″) parts—form the fundamental basis for calculating the total impedance (Z) at each applied frequency. By evaluating the ratio of these components, the complex impedance spectrum can be constructed.

These data can be represented through different plotting methods, among which the Nyquist plot is one of the most widely used. In the current analysis, such plots effectively reveal both the bulk conductive characteristics and the influence of grain or interfacial boundaries on ion transport behaviour.

An EIS setup for SPEs as shown in Figure 2.5 typically involves:

a) Potentiostat/Galvanostat: This is the control unit. It precisely maintains the desired DC electrochemical conditions (e.g., open circuit voltage) across the cell while simultaneously superimposing the small AC signal.

b) Frequency Response Analyzer (FRA): Frequently combined with the potentiostat, the FRA generates the required AC frequencies and precisely records the amplitude and phase shift of the corresponding response signal. It then calculates the impedance values at each frequency.

c) Electrochemical Cell: This is where the SPE sample resides. For measuring bulk ionic conductivity of SPEs, a symmetric two-electrode cell is the standard [32]. The SPE film is sandwiched between two identical, inert, blocking electrodes (e.g., stainless steel, gold, or platinum) [33]. "Blocking" means these electrodes prevent faradaic (charge-transfer) reactions, ensuring that the measured impedance primarily reflects ion transport within the electrolyte and ion accumulation at the interfaces.

d) Computer and Software: These elements form the command centre, controlling the entire experiment, acquiring the vast dataset of impedance vs. frequency, and performing the complex mathematical operations (like the Fourier Transform if the FRA uses time-domain measurements) to generate the final plots and allow for data fitting.

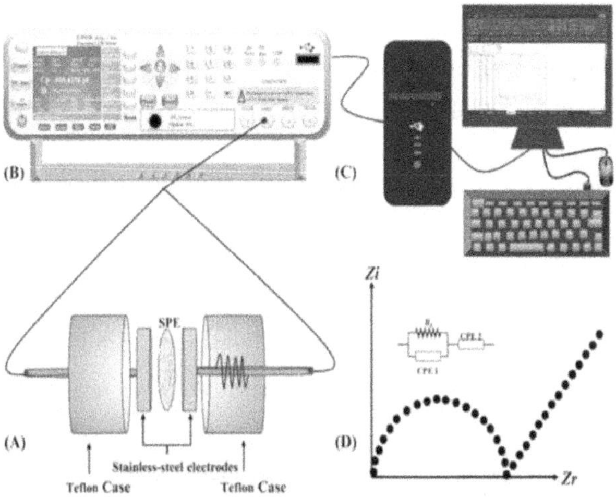

Figure 2.5 Setup and cell configuration for the EIS measurement, (**A**) cell arrangement, (**B**) LCR meter, (**C**) computer, (**D**) Nyquist plot.

e) Temperature Control: Since ion transport in polymers is highly sensitive to temperature, the electrochemical cell is typically placed inside a precisely controlled oven or climate chamber to ensure measurements are taken at stable, known temperatures.

2.2.3.2 Sample Preparation of SPEs for EIS analysis:

Accurate EIS data for SPEs heavily relies on meticulous sample preparation:

- **Uniform Film Fabrication:** The SPE should be prepared as a thin, uniform film (e.g., via solution casting, hot pressing or electro-spinning). Any non-uniformity or defects (like pinholes) can lead to short circuits or inaccurate measurements.
- **Well-Defined Geometry:** Accurate measurement of the electrode's active surface area and the thickness of the SPE film is essential. These values are crucial for converting measured resistance into absolute ionic conductivity.
- **Good Electrical Contact:** Achieving excellent physical contact between the SPE film and the blocking electrodes is paramount. Poor contact can introduce significant "contact resistance," which can mask the true bulk electrolyte resistance. This is often achieved by applying gentle, uniform pressure across the cell using spring-loaded setups or standard coin cell assemblies [34].
- **Environmental Control:** The SPE samples are often hygroscopic (absorb moisture) and sensitive to atmospheric oxygen [35]. Therefore, sample preparation and cell assembly are frequently performed in a dry room or glovebox (e.g., argon-filled) to prevent contamination that could drastically alter the electrolyte's properties.

The electrical characteristics of the material were examined using AC impedance spectroscopy performed between conductive electrodes. This technique not only reveals impedance behaviour but also facilitates the extraction of related parameters such as dielectric permittivity, electric modulus, capacitance, and dissipation factor (tan δ). By applying a low-amplitude alternating signal across the sample, the response is recorded on the monitoring system with minimal disturbance to the electrochemical setup, thus enhancing measurement accuracy across the frequency spectrum.

In this work, AC impedance analysis was employed to investigate the ion transport and dielectric behaviour of polymer electrolyte films. Measurements were conducted across a temperature range of 303–353 K and a frequency window spanning from 1 Hz to 1 MHz, utilizing a computer-interfaced HIOKI 3532 LCR meter. A visual depiction of the instrument setup is provided in Figure 2.6.

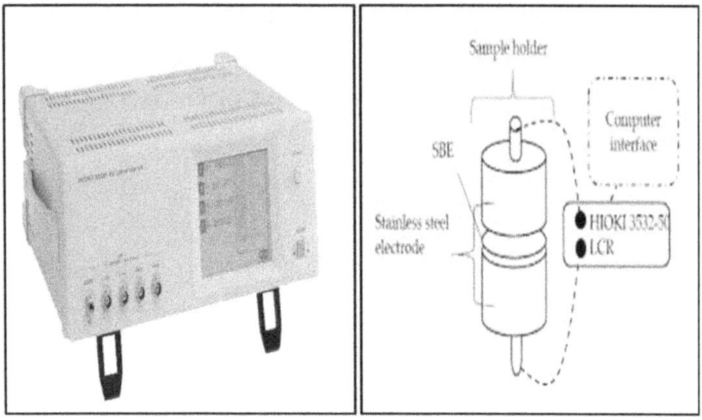

Figure 2.6 Photograph and schematic diagram of LCR meter.

EIS is a profoundly informative technique for SPE characterization. By applying tiny AC signals over a wide frequency range and meticulously analyzing the system's response, it provides a powerful means to quantify the intrinsic ionic conductivity of the polymer, understand interfacial phenomena, and identify bottlenecks to ion transport, thereby guiding the rational design and optimization of SPEs for high-performance electrochemical devices.

2.2.4 Conductivity studies

2.2.4.1. Nyquist Plot (Cole-Cole plot)

The Nyquist or complex impedance plot is a widely adopted graphical representation for analysing electrochemical impedance spectra. In this format, the real component of impedance (Z') is plotted against its imaginary counterpart (Z'') across a range of frequencies, generating a profile that typically comprises a semicircular arc at higher frequencies and a slanted tail at lower frequencies. Every point on the curve corresponds to the system's response at a specific frequency, where higher frequencies are grouped on the left side of the plot, while lower frequencies are positioned toward the right.

The high-frequency arc corresponds to the material's bulk response, while the lower frequency slope reflects interfacial phenomena. From this arc, the bulk resistance (R_b) of the polymer film can be determined by extending the semicircle to intersect the real axis, providing a clear estimation of the material's intrinsic resistive properties.

Figure 2.7 illustrate the theoretical impedance arc. At the leftmost (high-frequency) edge of the semicircle, the impedance is governed solely by R_b. As frequency decreases, additional impedance arises from polarization effects, causing the total resistance to approach $R_b + R_p$ at the low-frequency limit.

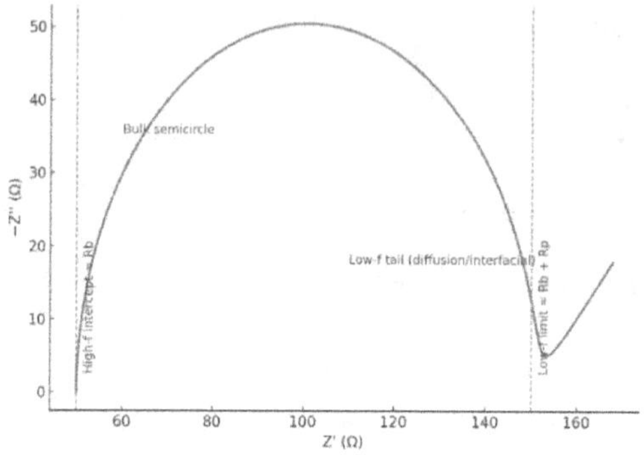

Figure 2.7 Theoretical Nyquist plot for SPE.

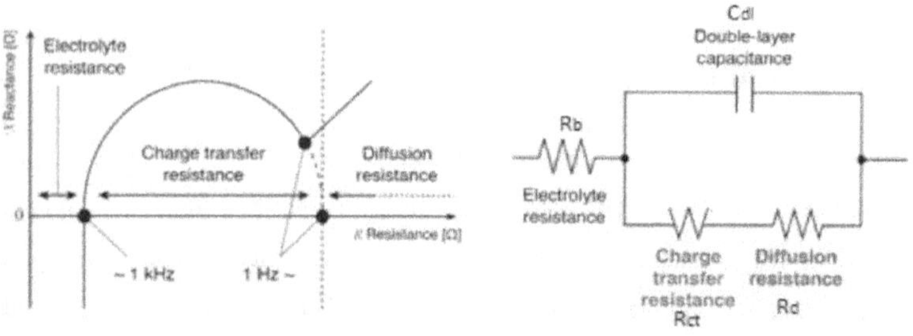

Figure 2.8 Nyquist plot and its equivalent circuit.

The bulk resistance (R_b) of the polymer film was extracted by fitting the experimental Cole–Cole plot using Z-View software. A representative plot along with its corresponding equivalent circuit is presented in Figure 2.8. The shape and features of the Nyquist plot vary notably among different polymer electrolyte systems, influenced by factors such as polymer architecture, degree of amorphous content, nature of charge carriers, and ion mobility within the matrix [36].

Interpretation of Nyquist plot (Cole-Cole plot):

High-Frequency Intercept on X-axis (R_b): The most crucial feature for SPEs. The position where the curve intersects the real axis at higher frequencies corresponds to the **bulk resistance (R_b)** of the solid polymer electrolyte. This denotes the intrinsic resistance of the material to ionic movement. A smaller intercept signifies lower bulk resistance and thus higher ionic conductivity. The **ionic conductivity (σ)** is determined through the equation:

$$\sigma = d/R_b.A \qquad \text{......... 2.12}$$

Where, d denotes the thickness of the SPE film (cm), *A* represents the active electrode area (cm^2), and *Rb* corresponds to the bulk resistance (Ω).

Depressed Semicircle (Mid-Frequencies): Following the high-frequency intercept, a depressed (flattened) semicircle is often observed. This characteristic is generally ascribed to the charge transfer resistance (R_{ct}) at the electrode–electrolyte interface, together with the geometric or double-layer capacitance (C_{dl}). The appearance of a "depressed" arc, rather than a perfect semicircle, typically reflects non-ideal capacitive behavior, which is often represented by a Constant Phase Element (CPE) to account for factors such as surface irregularities or uneven current distribution [37]. A smaller semicircle indicates better interfacial contact and lower charge transfer resistance.

Low-Frequency Spike (Blocking Electrodes): At very low frequencies, the plot typically shows a steep vertical or nearly vertical line (a "spike") parallel to the imaginary axis. This is characteristic of **blocking electrodes**, where ions accumulate at the interface without undergoing faradic reactions, leading to a large capacitive response.

2.2.4.2 Conductance spectra

The conductance spectrum, represented by a plot of log conductivity (log σ) versus log angular frequency (log ω), serves as a powerful tool for evaluating ionic dynamics in polymer electrolytes. This spectrum reveals critical phenomena such as space charge polarization, bulk relaxation, and the hopping motion of mobile ions. Each frequency domain corresponds to different ion transport mechanisms within the polymer matrix.

The DC conductivity (σ_0) is determined by extending the low-frequency plateau region to zero frequency, representing the long-range ionic movement. The observed conductivity is determined by the ionic concentration as well as the mobility of charge carriers in the electrolyte.

The AC conductivity, $\sigma(\omega)$, shows a frequency-dependent behaviour consistent with Jonscher's universal power law:

$$\sigma(\omega) = \sigma_0 + A\omega^{\alpha} \quad \text{........... 2.13}$$

Here, σ_0 represents the DC conductivity, A corresponds to a temperature-dependent pre-exponential constant, and α ($0 < \alpha < 1$) defines the dispersion characteristics arising from localized ion hopping. This model effectively captures the relationship between conductivity and the dynamic motion of ions embedded in a disordered matrix [38].

The conductivity spectra exhibit three distinct regions:

Low-Frequency Plateau:

The flat region signifies the DC ionic conductivity (σ_0) of the bulk electrolyte, indicating long-range ion migration [39]. The extrapolated value at zero frequency provides the conductivity relevant for device performance.

Mid-Frequency Dispersion:

As frequency increases from the plateau, the conductance often starts to rise, showing a dispersion [40]. This region typically describes the increase in Conductivity with frequency, conforming to Jonscher's power law

$$\sigma(\omega) = \sigma_0 + A\omega^{\alpha} \quad \text{.......... 2.14}$$

where A is a temperature-dependent constant and α ($0 < \alpha < 1$) reflects ion hopping dynamics [41]. This region indicates localized motion of charge carriers, influenced by polymer chain flexibility and free volume within the polymer matrix.

At elevated frequencies, mobile ions engage in more localized or constrained hopping events, enhancing the AC conductivity despite the absence of long-range transport. The parameter n, characterizes the nature of these movements. An n value approaching 1 implies nearly ideal, uncorrelated ion motion, whereas lower values indicate a greater degree of disorder or interaction among hopping events within the polymer matrix.

High-Frequency Spike:

At elevated frequencies, a sharp rise in $\sigma(\omega)$ is observed, associated with dielectric relaxation due to dipolar reorientation or constrained ionic oscillations. Minor contributions from instrumental effects may also appear here.

This analysis provides a nuanced view of conduction behaviour, distinguishing between mobile ion pathways and relaxation effects—complementing insights gained from structural (PXRD) and chemical (FTIR) studies.

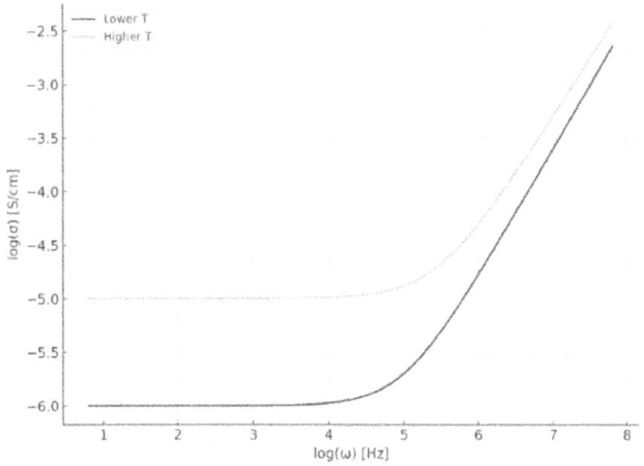

Figure 2.9 Conductance spectra of solid polymer electrolyte films measured at different temperatures.

The graph highlights three distinct regions: (i) the low-frequency DC plateau (σ_0), (ii) the intermediate-frequency dispersion described by Jonscher's law, and (iii) the high-frequency dielectric relaxation region.

2.2.4.3 Temperature dependent conductivity

Temperature is a critical factor influencing ionic conductivity, especially in solid polymer electrolytes (SPEs), where ion transport is closely associated with the dynamic behaviour of the polymer matrix. To elucidate this intricate relationship, several empirical models have been developed, each offering a distinct perspective on the underlying ion transport phenomena.

In solid polymer electrolytes, the temperature-dependent conductivity often exhibits behaviours that can be described by **Arrhenius theory**

Arrhenius Theory:

The Arrhenius theory, originally formulated to describe reaction rates, extends to ionic conduction by postulating that ions hop between available sites by surmounting an energy barrier. In the context of SPEs, this implies that the mobile ions (both cations and anions, though lithium ions are typically the primary charge carriers in Li-ion applications) overcome an activation energy (E_a) to move from one coordination site to another within the polymer matrix.

The fundamental Arrhenius equation adapted to describe the electrical conductivity (σ) in SPEs is expressed as:

$$\sigma = Ae^{(-Ea/RT)} \quad \text{..........2.15}$$

Where:

σ (Ionic Conductivity): In SPEs, k from the general Arrhenius equation is replaced by σ, representing the material's ability to conduct ions. This is a measure of how efficiently charge carriers (ions from the dissolved salt) move through the polymer medium.

A (Pre-exponential Factor): For SPEs, the pre-exponential factor A is not merely a collision frequency. Instead, It involves parameters such as the concentration of mobile ions, their intrinsic hopping frequency, and the availability of hopping sites within the polymer framework. A higher A value suggests more charge carriers or more frequent successful hopping attempts.

Ea (Activation Energy): This is perhaps the most critical parameter in the context of SPEs. It represents the minimum energy barrier that ions must overcome to move from one coordination site to another within the polymer matrix. Uniquely for SPEs, this activation energy is intrinsically associated with the segmental mobility of polymer chains. Ions in SPEs are often solvated by the polymer segments (e.g., ether oxygens in polyethylene oxide, PEO, coordinating with Li^+ ions). For an ion to hop, the surrounding polymer segments must locally rearrange to open up a pathway. Thus, the Ea reflects not just the energy for ion migration but also the energy required for this cooperative polymer segmental relaxation. A lower Ea indicates easier ion transport and chain flexibility.

R (Universal Gas Constant) and T (Absolute Temperature): These retain their standard thermodynamic meanings, dictating the thermal energy available to facilitate ion transport.

For SPEs with relatively rigid backbones or limited amorphous content, the ion transport mechanism often resembles that in crystalline or glassy solids, where hopping between fixed sites dominates. When lnσ is plotted against 1/T (an Arrhenius plot), a linear relationship indicates that the ionic conduction follows Arrhenius behaviour. The slope of this linear plot is equal to **−Ea/R**, allowing for the direct calculation of the activation energy, **Ea**. A lower activation energy indicates easier ion migration, often associated with increased polymer chain flexibility, reduced crystallinity, or the incorporation of suitable plasticizers or fillers.

However, many SPEs, particularly those with significant amorphous content, exhibit non-Arrhenius behaviour at temperatures near or below their glass transition temperature (Tg) . The deviation from linearity in the Arrhenius plot indicates that ionic transport is not exclusively controlled by straightforward activated hopping, but is closely associated with the segmental dynamics of polymer chains and the presence of available free volume.

The Arrhenius model is generally applicable to rigid solid ion conductors or to polymer electrolytes at temperatures significantly above their glass transition temperature (Tg), where the polymer chains possess sufficient thermal energy for extensive segmental motion, and ion hopping is less constrained by the polymer's viscous flow. It simplifies the ion transport to a thermally activated hopping process over a fixed energy barrier.

2.2.4.4 Dielectric Study

The dielectric behaviour of a material reflects its ability to both store energy through polarization and dissipate energy via dielectric loss when subjected to an external electric field, making it a crucial factor in the study of solid polymer electrolytes, capacitors, and other electrochemical systems.

The dielectric property of a material mainly indicates its short-range electrical conduction when exposed to an applied electric field. When such a field is applied, it causes displacement of charges within the medium, leading to charge accumulation at interfaces and the formation of dipoles with associated moments. The relaxation of these dipoles originates from the reorientation of molecular dipoles within the structure. At its core, this behaviour results from the interaction of the applied electric field with the inherent dipole moments and charge carriers within the material.

The dielectric constant (ε') of solid polymer electrolytes represents their capacity to store electrical energy via polarization when subjected to an external field. It corresponds to the real component of the complex permittivity (ε*), expressed as:

$$\varepsilon^*(\omega) = \varepsilon'(\omega) - j\varepsilon''(\omega) \quad \text{......2.16}$$

Here, **ε′** denotes the dielectric constant, while **ε″** represents the dielectric loss. The dielectric constant is expressed as:

$$\varepsilon'(\omega) = \frac{C\,d}{\varepsilon_0 A} \quad \text{.......2.17}$$

with *C* being the capacitance, *d* the sample thickness, *A* the electrode area, and ε_o the permittivity of free space.

At low frequencies, ε′ typically exhibits elevated values as a result of electrode polarization and the build-up of space charges at the electrode–electrolyte interfaces. As frequency increases, ε′ decreases sharply because dipoles and mobile ions cannot align rapidly with the alternating field, revealing the intrinsic bulk dielectric response of the polymer matrix [42]. A higher dielectric constant generally facilitates greater ion dissociation, thereby enhancing ionic conductivity.

The dielectric loss (ε") represents the energy dissipated within the electrolyte material due to ion migration and dipolar relaxation, and is given by:

$$\varepsilon''(\omega) = \frac{\sigma(\omega)}{\varepsilon_0\,\omega} \quad \text{...... 2.18}$$

where $\sigma(\omega)$ is the AC conductivity and ω the angular frequency.

At low frequencies, elevated ε″ values arise from interfacial polarization and ion migration losses, while at higher frequencies, ε″ decreases as charge carriers fail to follow the rapidly oscillating field. Peaks in ε″ spectra often signify relaxation processes associated with localized ionic hopping and polymer chain segmental motion [43].

Thus, analysis of ε′ and ε″ provides direct insights into charge storage ability, interfacial polarization, dielectric relaxation, and energy dissipation, thereby linking dielectric behaviour with ionic transport efficiency in SPEs.

To mitigate electrode polarization effects and more accurately capture bulk relaxation, electric modulus analysis (M′, M″) is often employed; the imaginary modulus peak corresponds to conductivity relaxation processes [43]. Dielectric relaxation times derived from loss tangent spectra (tan δ) indicate dipolar and ionic mobility shifts with temperature and structural modifications, showing non-Debye behaviour [44]. Collectively, these dielectric parameters correlate strongly with ionic transport and segmental chain motion, enhancing the design and optimization of SPEs for efficient solid-state devices.

2.2.4.5 Electric modulus

The electric modulus (M*) formalism—defined as the reciprocal of the complex permittivity—serves as an effective approach for examining the dielectric response of solid polymer electrolytes (SPEs), especially in distinguishing bulk relaxation from electrode polarization effects. It is expressed as:

$$M^* = \frac{1}{\varepsilon^*} = M' + jM'' \quad \text{...... 2.19}$$

where,

$$M' = \frac{\varepsilon'}{\varepsilon'^2 + \varepsilon''^2}, \quad M'' = \frac{\varepsilon''}{\varepsilon'^2 + \varepsilon''^2}$$

The real part M′ represents the elastic energy storage capability of the medium, while the imaginary component M″ corresponds to relaxation processes linked to ionic motion.

In typical spectra, M′ remains near zero at lower frequencies, illustrating the suppression of interfacial polarization, and increases at higher frequencies due to ionic or dipolar relaxation onset, reflecting bulk relaxation. The peak of M″ directly marks the conductivity relaxation frequency, enabling estimation of ionic mobility and characteristic relaxation times [45].

Unlike direct dielectric permittivity analysis, which is often dominated by interfacial polarization at low frequencies, the modulus approach effectively suppresses such contributions and emphasizes bulk relaxation and ionic mobility.

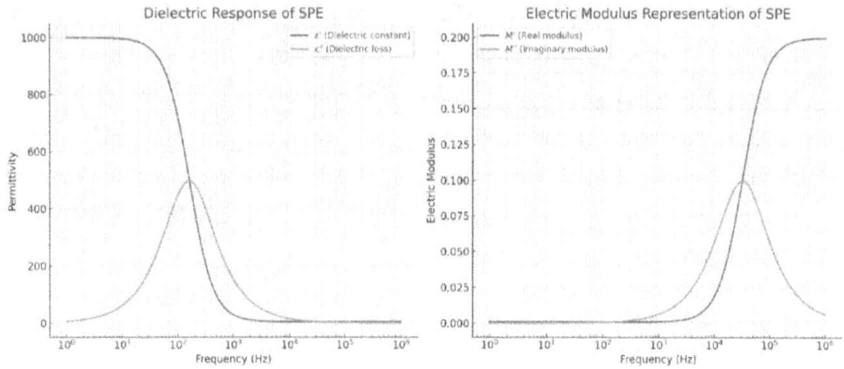

Figure 2.9.1 Dielectric response and electric modulus of SPEs. The first plot shows the dielectric constant (ε') and dielectric loss (ε'') versus frequency, while the second plot represents the real (M') and imaginary (M'') partsof the electric modulus.

Recent dielectric investigations of SPEs report that the presence and positioning of these M'' peaks can shift depending on amorphous content, salt concentration, and temperature, highlighting how polymer segmental dynamics influence ion transport. For example, Awang et al. (2022) observed a pronounced M'' peak in starch-based SPEs, consistent with non-Debye conductivity relaxation. Similarly, Zhao et al. (2022) identified relaxation peaks in M'' spectra of Mg^{2+}-doped PEO/PVdF-HFP blends, linking frequency shifts to reduced amorphous phase mobility [46]. These modulus insights are integral to understanding and optimizing the transport behaviour of SPEs.

2.2.4.6 Tangent Studies

The loss tangent (tan δ) is an important dielectric parameter that evaluates the ratio between energy storage and energy dissipation in solid polymer electrolytes (SPEs). The loss **tangent (tan δ)** is defined as the ratio of the dielectric loss to the dielectric constant

$$tan\,\delta = \frac{\varepsilon_r''}{\varepsilon_r'} \quad \text{...... 2.20}$$

Where tan δ provides insight into how efficiently a material stores electrical energy relative to how much is lost as heat.

In SPEs, tan δ typically exhibits strong frequency dependence. At low frequencies, higher tan δ values are often observed due to electrode polarization and interfacial charge accumulation. As frequency increases, the dipoles and mobile ions can no longer keep pace with the rapidly oscillating field, leading to a reduction in tan δ [47].

The peak of the loss tangent is defined by the following relation for the highest dielectric loss:

$$\omega\tau = 1 \quad \text{....... 2.21}$$

where τ denotes applied electric field's angular frequency and ω is the relaxation duration. This reduction in relaxation time increased the ionic conductivity.

The peak in tan δ spectra corresponds to the material's characteristic relaxation processes, such as ionic hopping or dipole reorientation; these peaks shift toward higher frequencies with increasing temperature or filler content, indicating enhanced polymer segmental mobility and conductivity [48].

2.2.4.7 Argand Plot Studies

The Argand plot, commonly referred to as the Cole-Cole plot, is a graphical tool employed to study the electrical characteristics of solid polymer electrolytes (SPEs). It represents a graph of the imaginary component of the electrical modulus (M") versus its real component (M'), both obtained from impedance analysis. This helps us understand how electric charges move and relax inside the polymer material.

In ideal cases, the plot looks like a semicircle. The size and position of the arc give information about how fast dipoles in the polymer reorient and how strong their response is. For SPEs, the Argand plot often deviates from a perfect semicircle, showing depressed or distorted arcs due to non-ideal dielectric relaxation caused by polymer chain segmental motion, ion–dipole interactions, and structural disorder. The extent of distortion indicates the non-uniformity of the polymer matrix and the intricate nature of ionic transport routes. At high frequencies, the arc tends to shrink, highlighting fast dipolar relaxation, while at low frequencies, tail-like extensions may emerge, attributed to electrode polarization and space charge accumulation.

2.3 References

[1] Gong, X.; Wang, J.; Zhong, L.; Qi, G.; Liu, F.; Pan, Y.; Yang, F.; Wang, X.; Li, J.; Li, L.; Liu, C.; Yu, D. Recent advances on cellulose-based solid polymer electrolytes. *Ind. Chem. Mater.* **2025**, *3*, 31–48.

[2] Mohan, J.; Vijayachamundeeswari, S.P. The significance of fillers in composite polymer electrolytes for optimizing lithium battery. *Ionics* **2024**, *30*, 647–675.

[3] Teo, L.P.; Buraidah, M.H.; Arof, A.K. Development on solid polymer electrolytes for electrochemical devices. *Molecules* **2021**, *26*, 6499.

[4] Augustine, K. Fabricating and analyzing liquid and polymer electrolytes for sodium-ion batteries. *World J. Adv. Res. Rev.* **2025**, *26*, 1148–1188.

[5] Toolan, D.T.W.; Isakova, A.; Hodgkinson, R.; Reeves-McLaren, N.; Hammond, O.S.; Edler, K.J.; Briscoe, W.H.; Arnold, T.; Gough, T.; Topham, P.D.; Howse, J.R. Insights into the influence of solvent polarity on the crystallization of poly(ethylene oxide) spin-coated thin films via in situ grazing incidence wide-angle X-ray scattering. *Macromolecules* **2016**, *49*, 4579–4586.

[6] Strawhecker, K.; Kumar, S.K. The critical role of solvent evaporation on the roughness of spin-cast polymer films. *Macromolecules* **2001**, *34*, 14.

[7] Murawski, T.T.; Żołek-Tryznowska, Z.; Szałapak, J. Impact of substrate type on the properties of cast biodegradable starch-based films. *Processes* **2025**, *13*, 1197.

[8] Han, P.; Zhu, Y.; Liu, J. An all-solid-state lithium ion battery electrolyte membrane fabricated by hot-pressing method. *J. Power Sources* **2015**, *284*, 1–9.

[9] Ahmed, F.; Chen, A.; Altoé, M.V.P.; Liu, G. Argyrodite-Li6PS5Cl/polymer-based highly conductive composite electrolyte for all-solid-state batteries. *ACS Appl. Energy Mater.* **2024**, *7*, 1842–1853.

[10] Reddy, V.S.; Tian, Y.; Zhang, C.; Ye, Z.; Roy, K.; Chinnappan, A.; Ramakrishna, S.; Liu, W.; Ghosh, R. A review on electrospun nanofibers based advanced applications: From health care to energy devices. *Polymers* **2021**, *13*, 3746.

[11] Liu, Z.; Ramakrishna, S.; Ahmed, I.; Rudd, C.; Liu, X. Rheological, surface tension and conductivity insights on the electrospinnability of poly(lactic-co-glycolic acid)-hyaluronic acid solutions and their correlations with the nanofiber morphological characteristics. *Polymers* **2022**, *14*, 4411.

[12] Al-Abduljabbar, A.; Farooq, I. Electrospun polymer nanofibers: Processing, properties, and applications. *Polymers* **2023**, *15*, 65.

[13] Haider, A.; Haider, S.; Kang, I.-K. A comprehensive review summarizing the effect of electrospinning parameters and potential applications of nanofibers in biomedical and biotechnology. *Arab. J. Chem.* **2018**, *11*, 1165–1188.

[14] Huang, F.; Singer, G.; Zhou, Y.; Wang, C.H. Creating ionic pathways in solid-state polymer electrolyte by using PVA-coated carbon nanofibers. *Compos. Sci. Technol.* **2021**, *207*, 108710.

[15] Jeanne-Brou, R.; Charvin, N.; De Moor, G.; Devaux, D. Ionic conductivity of solid polymer electrolytes depending on elongation. *Electrochim. Acta* **2023**, *469*, 143253.

[16] Sun, M.; Zeng, Z.; Zhong, W.; Xie, J. In-situ polymerization methods for polymer-based solid-state lithium batteries. *Batteries Supercaps* **2022**, *5*, 12.

[17]. Stan, C. V.; Beavers, C. M.; Kunz, M.; Tamura, N. X-Ray Diffraction under Extreme Conditions at the Advanced Light Source. *Quantum Beam Sci.* **2018**, *2* (1), 4.

[18] Varlashkin, P. Approaches to quantification of amorphous content in crystalline drug substance by powder X-ray diffraction. *Am. Pharm. Rev.* **2011**, *14*, 1.

[19] Teo, L.P.; Buraidah, M.H.; Arof, A.K. Development on solid polymer electrolytes for electrochemical devices. *Molecules* **2021**, *26*, 6499.

[20] Ningappa, N.G.; Reddy, A.K.M.; Zaghib, K. Advanced polymer electrolytes in solid-state batteries. *Batteries* **2024**, *10*, 454.

[21] Chattopadhyay, J.; Pathak, T.S.; Santos, D.M.F. Applications of polymer electrolytes in lithium-ion batteries: A review. *Polymers* **2023**, *15*, 3907.

[22] Luo, R.; Jiang, H.; Du, B.; Zhou, S.; Zhu, Y. Preparation and application of solid polymer electrolyte based on deep eutectic solvent. *AIP Adv.* **2019**, *9*, 035341.

[23] Hoang, V.P.H.; So, S.; Hur, J. Inorganic fillers in composite gel polymer electrolytes for high-performance lithium and non-lithium polymer batteries. *Nanomaterials* **2021**, *11*, 614.

[24] Koczoń, P.; Hołaj-Krzak, J.T.; Palani, B.K.; Bolewski, T.; Dąbrowski, J.; Bartyzel, B.J.; Gruczyńska-Sękowska, E. The analytical possibilities of FT-IR spectroscopy powered by vibrating molecules. *Int. J. Mol. Sci.* **2023**, *24*, 1013.

[25] Griffiths, P.; de Hasseth, J.A. Fourier Transform Infrared Spectrometry, 2nd ed.; Wiley-Blackwell: Hoboken, NJ, USA, **2007**; ISBN 978-0-471-19404-0.

[26] Fadlelmoula, A.; Pinho, D.; Carvalho, V.H.; Catarino, S.O.; Minas, G. Fourier transform infrared (FTIR) spectroscopy to analyse human blood over the last 20 years: A review towards lab-on-a-chip devices. *Micromachines* **2022**, *13*, 187.

[27] Ciolacu, D.E.; Ciolacu, F.; Popa, V.I. Amorphous cellulose—Structure and characterization. *Cellulose Chem. Technol.* **2011**, *45*, 13–21.

[28] da Silva, G.M.G.; Faia, P.M.; Mendes, S.R.; Araújo, E.S. A review of impedance spectroscopy technique: Applications, modelling, and case study of relative humidity sensors development. *Appl. Sci.* **2024**, *14*, 5754.

[29] Goel, H.; Kumar, R.; Kant, R. Modular theory for EIS response of electron transfer coupled with electric double layer on rough and heterogeneous electrode. *Electrochim. Acta* **2023**, *468*, 143125.

[30] Zhang, L.; Dai, Y.; Li, C.; Dang, Y.; Zheng, R.; Wang, Z.; Wang, Y.; Cui, Y.; Arandiyan, H.; Shao, Z.; Sun, H.; Zhuang, Q.; Liu, Y. Recent advances in electrochemical impedance spectroscopy for solid-state batteries. *Energy Storage Mater.* **2024**, *69*, 103378.

[31] Takahashi, T.; Kuwabara, K.; Shibata, M. Conductivities of Na^+ Ion Conductors Based on NASICON. *Solid State Ion.* **1980,** *1*, 163–175.

[32] Biancolli, A.L.G.; Konovalova, A.; Santiago, E.I.; Holdcroft, S. Measuring the ionic conductivity of solid polymer electrolyte powders. *Int. J. Electrochem. Sci.* **2023**, *18*, 100288.

[33] Abdul Halim, S.I.; Chan, C.H.; Apotheker, J. Basics of teaching electrochemical impedance spectroscopy of electrolytes for ion-rechargeable batteries – part 1: A good practice

on estimation of bulk resistance of solid polymer electrolytes. *Chem. Teacher Int.* **2021**, *3*, 20200011.

[34] Isaac, J.A.; Mangani, L.R.; Devaux, D.; Bouchet, R. Electrochemical impedance spectroscopy of PEO-LATP model multilayers: Ionic charge transport and transfer. *ACS Appl. Mater. Interfaces* **2022**, *14*, 13158–13168.

[35] Subhani, T.; Khademolqorani, S.; Banitaba, S.N.; Ramadan, M.; Khaliq, A.; Chaudhry, I.A.; Osman, A.I. Advancements in battery materials: Bio-based and mineral fillers for next-generation solid polymer electrolytes. *ACS Appl. Mater. Interfaces* **2024**, *16*, 63089–63108.

[36] Elizalde, M.; Trifonov, A.; Karimi, A.; Carrete, J. Exploring structure–property relationships in PEO-based polymer electrolytes: Effect of polymer architecture and filler on ion transport and impedance response. *Electrochim. Acta* **2021**, *388*, 138657.

[37] Zhyhailo, M.; Yevchuk, I.; Ivashchyshyn, F.; Demchyna, O.; Chabecki, P.; Babkina, N.; Shantaliy, T. Modeling of electrochemical impedance of fuel cell based on novel nanocomposite membrane. *Energies* **2024**, *17*, 2754.

[38] Jonscher, A.K. The "universal" dielectric response. *Nature* **1977**, *267*, 673–679.

[39] Sivakumaran, A.; Surendran, V.; Butler, S.; Reid, S.; Thangadurai, V. Sodium ion conductivities in Na2O–Sm2O3–SiO2 ceramics. *EES Batteries* **2025**, *1*, 287–297.

[40] Dam, T.; Thomas, N.K.R.; Katiyar, R.S. Observation of ionic transport and ion-coordinated segmental motions in composite (polymer-salt-clay) solid polymer electrolyte. *Ionics* **2014**, *21*, 2.

[41] Jonscher, A.K. The "universal" dielectric response. *Nature* **1977**, *267*, 673.

[42] Vinodkumar, V.; Kavitha, V.; Kumar, M.; Viswanathan, B. Structural, surface morphological and dielectric studies of guanidinium salt incorporated PEO–PVP solid polymer electrolytes. *Heliyon* **2023**, *9*, e10732.

[43] Tripathi, N.; Shukla, A.; Thakur, A.K.; Marx, D.T. Dielectric modulus and conductivity scaling approach to the analysis of ion transport in solid polymer electrolytes. *Polym. Eng. Sci.* **2020**, *60*, 297–305.

[44] Pradeep, R.; Siva, V.; Anandha Jothi, M.; Murugan, A.; Shameem, A.; Sanjana, S.; Hemnath, E.; Al-Sehemi, A.G. Structural, surface morphological and dielectric studies of

guanidinium salt incorporated poly(ethylene oxide)/poly(vinyl pyrrolidone) solid polymer electrolytes. *Heliyon* **2024**, *10*, e23524.

[45] Awang, F.F.; Hassan, M.F.; Kamarudin, K.H. Study on ionic conductivity, dielectric and electrochemical properties of solid polymer electrolyte for battery applications. *Ionics* **2022**, *28*, 1249–1263.

[46] Money, B. K.; Hariharan, K.; Swenson, J. J. S. S. I. Relation between structural and conductivity relaxation in PEO and PEO based electrolytes. *Solid State Ionics* **2014**, *262*, 785-789.

[47] Sadiq, M.; Khan, A.; Raza, M.M.H.; Aalam, S.M.; Zulfequar, M.; Ali, J. Enhancement of electrochemical stability window and electrical properties of CNT-based PVA–PEG polymer blend composites. *ACS Omega* **2022**, *7*, 40116–40131.

[48] Verma, M.L.; Sahu, H.D. Ionic conductivity and dielectric behaviour of PEO-based silver ion conducting nanocomposite polymer electrolytes. *Ionics* **2015**, *21*, 3223–3231.

Chapter 3

Investigation of PVDF-HFP/PVP Polymer Blend Electrolytes Incorporating Lithium Iodide: Structural, Dielectric, And Electrochemical Characterization

3.1 Introduction

Poly (vinylidene fluoride-co-hexafluoropropylene) (PVDF-HFP) is widely chosen as a host polymer, mainly due to its excellent electrochemical stability, relatively high dielectric constant, and semi-crystalline character that supports ion transport. Due to its inherent crystallinity that limits segmental motion, thereby restricting ionic transport, polyvinylpyrrolidone (PVP) is incorporated owing to its amorphous structure, polar carbonyl functionality, and strong lithium-ion solvation capability. The incorporated of PVP into PVDF-HFP disrupts the crystalline domains, enhances amorphous content, and promotes salt dissociation, leading to improved ionic mobility.

Furthermore, the incorporation of lithium salt, Lithium iodide (LiI), with its relatively large anion and high solubility, has been employed to enhance ionic conductivity in the polymer matrices. The coordination of Li^+ ions with the electron-rich functional groups present in PVP and the fluorinated segments of PVDF-HFP establishes a conduction network where ions migrate via segmental polymer relaxation.

Structural and spectroscopic analyses, supported by dielectric and impedance measurements, provide insights into ion transport mechanisms, relaxation behaviour, and electrode–electrolyte interfacial stability.

The central aim of this chapter is to synthesize and study PVDF-HFP/PVP blend electrolytes containing lithium iodide, while emphasizing the relationship between polymer blending, salt inclusion, and the resulting electrochemical properties. The study highlights the structural modifications, dielectric, and electrochemical characteristics that support the development of efficient solid polymer electrolytes for advanced energy storage systems.

3.2 Preparation and Experimental Methods

Lithium ion-conducting copolymer electrolytes involving PVDF-HFP and PVP polymer blends with various compositions, PVDF-HFP (60 wt.%), PVP (40 wt.%), and LiI (5, 10, 15, and 20 wt.%) denoted as PPHLI5 wt.%, PPHLI10 wt.%, PPHLI15 wt.%, and PPHLI20 wt.%, respectively, were synthesised via the most widely adopted and straightforward Solution casting technique. To achieve a homogeneous and transparent solution, the required quantities of PVP, PVDF-HFP, and LiI were dissolved separately in DMF and continuously stirred for six hours. The solutions were then combined and continuously stirred for another six hours to induce gelation, followed by casting into Petri dishes. The resulting polymer films exhibited highly uniform surfaces, allowing for further characterisation and conductivity analysis. Figure 3.1 presents the preparation method of the polymer blend electrolyte film.

Structural and spectroscopic analyses were conducted using XRD and FTIR spectroscopy. Electrical characteristics, including ionic conductivity and dielectric behaviour, were evaluated via AC impedance analysis.

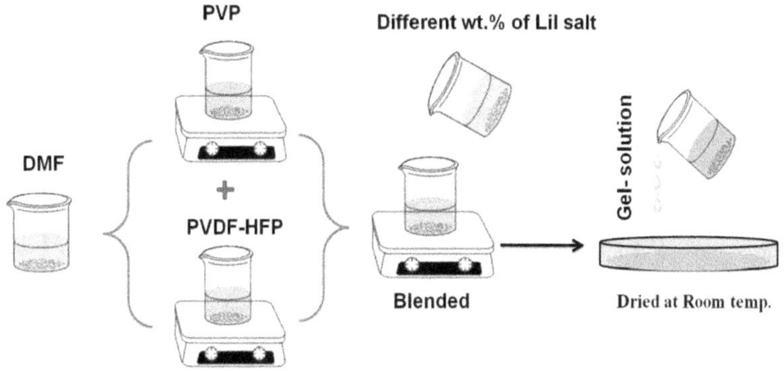

Figure 3.1 Solution Casting method for preparation of PVDF-HFP/ PVP/LiI solid polymer electrolytes.

3.3. Results and discussion

3.3.1 XRD analysis

The structural studies of different concentrations of lithium iodide (5, 10, 15, and 20 wt.%) mixed with specific concentrations of PVDF-HFP (60 wt.%), PVP (40 wt.%) blend denoted as PPHLI5 wt.%, PPHLI10 wt.%, PPHLI15 wt.%, and PPHLI20 wt.% is studied using XRD spectroscopy analysis and the recorded pattern is shown in Figure 3.2. The obtained XRD patterns reveal pure PVDF-HFP, PVP, and lithium iodide contents. From Figure 3.2, we can see that the typical peaks for PVDF-HFP show up at 20°, indicating that the PVDF polymer is present in a solid crystal form [1]. Further, characteristic peaks at 30° and 40° are detected. The peak at 30° shows the pure amorphous nature exhibited by the PVDF-HFP polymer. This indicates the reduction in the crystallinity of the SPEs with the inclusion of plasticisers and lithium iodide. Moreover, the XRD pattern does not reveal any characteristic peaks for lithium iodide. This indicates that lithium salts are effectively coordinated within the polymer matrix, confirming successful complex formation in the film. The increased amorphous nature of the material promotes greater flexibility in the polymer chains, which in turn facilitates improved ionic conductivity. Based on the results obtained, fixed complexation between the PVDF-HFP, the plasticiser, the PVP polymers, and the lithium iodide [2] in the resultant polymer film is confirmed.

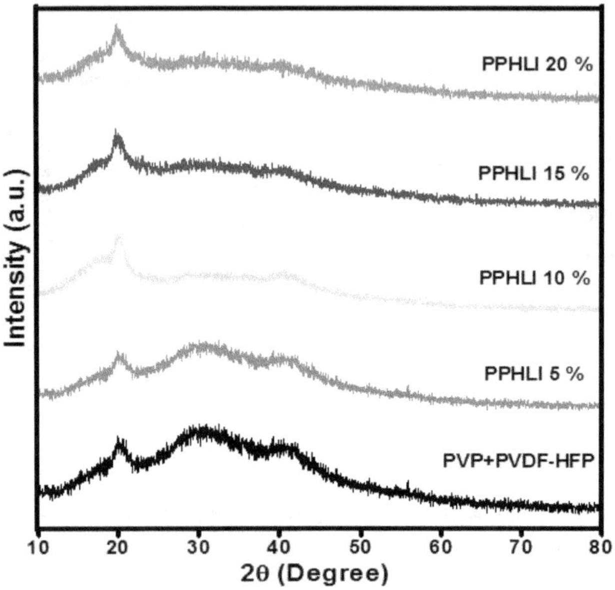

Figure 3.2 The XRD spectra of PVDF-HFP (60 wt.%): PVP (40 wt.%) blends prepared with 5, 10, 15, and 20 wt.% lithium iodides.

3.3.2 FTIR Analysis

Fourier-transform infrared spectrum (FTIR) is an essential record that provides detailed information throughout the polymer structure. The molecular bonding interactions within polymer electrolytes have been systematically analysed using FTIR, validating complex formation between polymers PVDF-HFP, PVP, and lithium iodide salt. The FTIR spectra obtained for the synthesised SPE films of PVDF-HFP, PVP, and LiI salts and its complexes are presented in Figure 3.3. FTIR transmittance spectra for the designated PPHLI5 wt.%, PPHLI10 wt.%, PPHLI15 wt.%, and PPHLI20 wt.% polymers are presented. The absorption bands at 1047 cm^{-1}, 1188 cm^{-1}, and 1656 cm^{-1} are assigned to CF_3 out-of-plane deformation in PVDF-HFP, antisymmetric stretching vibration of CF_2 groups in PVDF-HFP, and the C=O stretching vibration, respectively [3].Peaks observed at 1281 cm−1, 875 cm−1, and 813 cm−1 were attributed to α-phase formation at the PVDF-HFP crystalline peak [4]. The absorption band at 1422 cm^{-1} is associated with CH_2 wagging vibrations of the PVDF-HFP/PVP polymer backbone. The absorption band at 3469 cm^{-1} is attributed to O–H stretching in PVP, and incorporation of varying amounts of lithium iodide into PVDF-HFP/PVP brings about distinct changes in where the peak appears, as well as in its shape and intensity. The–CH3 asymmetric stretching in PVDF is assigned to the 2938 cm−1 peak [5]. Upon incorporation of lithium iodide into the synthesized polymer, the band typically observed near 760 cm^{-1} for crystalline PVDF-HFP shifted down to 741 $cm^{-1}$and exhibited reduced intensity, indicating the influence of increased salt concentration. These variations suggest enhanced amorphicity of the polymer matrix, arising from effective dispersion of the salt within the host system. Similarly, the band initially observed at 879 cm^{-1} was displaced to 877 cm^{-1}, further confirming salt–polymer complexation and the uniform distribution of lithium iodide throughout the polymeric network [6,7].

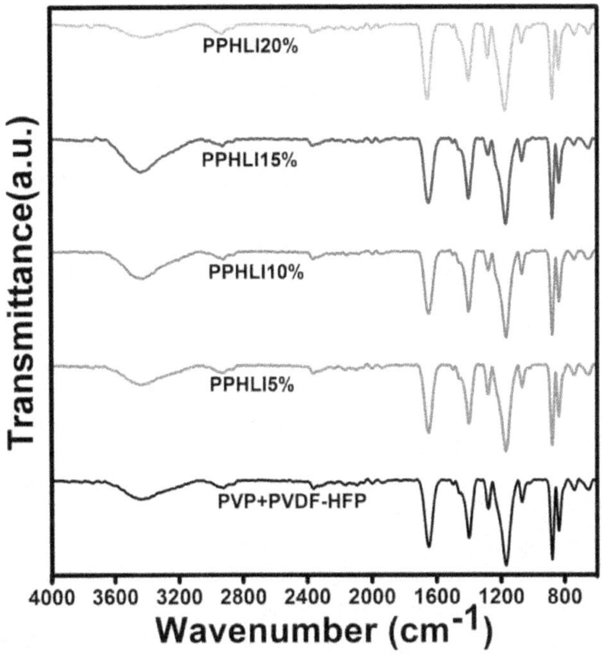

Figure 3.3 FTIR spectra of PVDF-HFP (60 wt.%): PVP (40 wt.%) blends with different concentrations of lithium iodide (5, 10, 15, and 20 wt.%).

A comprehensive FTIR peak assignment table for PVDF-HFP/PVP/LiI solid polymer electrolytes (SPEs) is presented in Table 3.1.

Table 3.1 Functional group vibration and assignment

Wavenumber (cm^{-1})	**Assignment**	**Origin / Interpretation**
3469 cm^{-1}	O–H stretching	PVP, Hydrogen bonded O-H stretching
2938 cm^{-1}	C–H stretching	CH_2 asymmetric stretching vibration of PVDF-HFP/PVP backbone.
1656 cm^{-1}	C=O stretching	PVP carbonyl group; strong Li^+–C=O coordination, evidence of salt–polymer complexation.
1422 cm^{-1}	CH_2 bending vibration	PVDF-HFP + PVP backbone vibrations.
1281 cm^{-1}	C–F stretching	PVDF-HFP backbone; modified by Li^+–F dipole interactions.
1188 cm^{-1}	CF_2 asymmetric stretching	PVDF-HFP skeletal vibrations; crystallinity-sensitive.
1047 cm^{-1}	C–N stretching	PVP pyrrolidone ring; shifts due to Li^+–C=O interaction.
875cm^{-1} 813 cm^{-1}	CH_2 rocking & CF_2 stretching	PVDF-HFP α phase formation of crystalline peak.

3.3.3 AC Impedance Analysis

3.3.3.1 Conductivity studies

AC impedance measurement serves as an excellent approach for probing the conduction processes and overall electrical performance of polymer electrolytes. The Nyquist plots or Cole-Cole plots for PVDF-HFP (60 wt.%): PVP (40 wt.%) with various concentrations of lithium iodide (5, 10, 15, and 20 wt.%) at ambient temperature are illustrated in Figure 3.4. For all the prepared solid polymer electrolytes, there exists a depressed semicircular portion. When increasing the salt content, a spike appears in the complex impedance representation. In comparison to a semicircle with a spike that corresponds to the resistance in a parallel connection and the capacitance in a series connection, the observed semicircles in this case are associated with equivalent electrical circuits where resistance and capacitance act in parallel. Notably, the semicircle obtained is relatively large. The linear region at lower frequencies reflects the influence of blocking electrodes, while this frequency range also represents the bulk response of the electrolyte. The real axis (Z') intercept obtained for the semicircle provides the electrolytic resistance (Rb). Bulk resistance was measured using Z-View software 3.5i version and an equivalent circuit fit.

Ionic conductivity was obtained using the equation

$$\sigma = l/R_bA \quad \text{......... 3.1}$$

where 'l' is the thickness of the polymer electrolyte film, 'R_b' is the bulk resistance found from the x-axis point in the complex impedance plot of the AC circuit, and 'A' is the surface area of the polymer film.

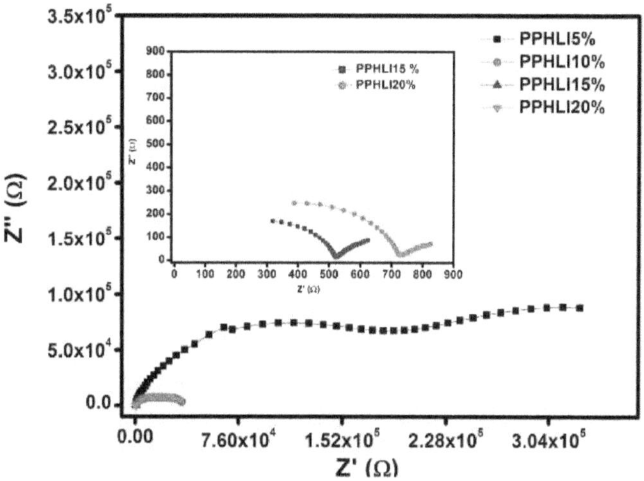

Figure 3.4 Cole–Cole plots of PPHLI5%, PPHLI10%, PPHLI15% and PPHLI20% SPE at room temperature.

As the lithium iodide concentration increases, the ionic conductivity through the solid polymer electrolyte increases. Figure 3.4 illustrates the better conductivity of PPHLI15 wt.% lithium iodide at room temperature when compared to other concentrations. The measured conductivity values are given in Table 3.2

Table 3.2 Conductivity of PVDF-HFP (60 wt.%): PVP (40 wt.%) blends with different compositions of lithium iodide (5, 10, 15, and 20 wt.%) at ambient temperature

Sample composition	Designated as	Conductivity $S{\cdot}cm^{-1}$
PVDF-HFP(60 wt.%): PVP(40 wt.%): LiI(5wt.%)	PPHLI 5%	1.2385×10^{-07}
PVDF-HFP(60 wt.%): PVP(40 wt.%): LiI(10wt.%)	PPHLI 10%	6.8678×10^{-07}
PVDF-HFP(60 wt.%): PVP(40 wt.%): LiI(15wt.%)	PPHLI 15%	7.504×10^{-05}
PVDF-HFP(60 wt.%): PVP(40 wt.%): LiI(20wt.%)	PPHLI 20%	2.837×10^{-05}

The prepared lithium iodide-doped PVDF-HFP/PVP polymeric electrolyte blend with composition PVDF-HFP (60 wt.%): PVP (40 wt.%): LiI (15wt.%) achieved its highest conductivity value of 7.504×10^{-5} $S{\cdot}cm^{-1}$.

The Impedance spectra analysis for the composition PVDF-HFP (60 wt.%): PVP (40 wt.%): LiI (15wt.%) at different temperatures reveal that the resistance of the electrolyte is found to decrease with an increase in temperature. This decrease in resistance appears due to enhanced ion mobility and the generation of more charge carriers with increasing temperature in the polymer electrolyte [8–10]. Figure 3.5 displays the impedance spectra of PVDF-HFP (60 wt.%): PVP (40 wt.%): LiI (15wt.%) at different temperatures. The semicircle diameter obtained is found to be reduced with an increase in thermal energy, which represents the improved bulk conductivity with rising temperature [2].

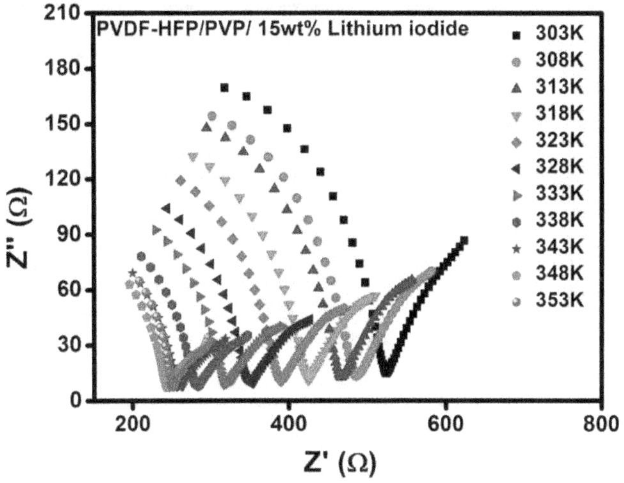

Figure 3.5 Impedance spectra for PVDF-HFP (60 wt.%): PVP (40 wt.%): LiI (15wt.%) heat-treated at different temperatures.

3.3.3.2 Measurements of Conductance Spectra

The conductance spectra of PVDF-HFP/PVP electrolytes with varying lithium iodide concentrations are illustrated in Figure 3.6. The figure shows the logarithmic variation of conductivity (σ) with angular frequency (ω) for all samples at room temperature. The spectra of the synthesized polymer electrolytes reveal two clear regions: a stable plateau at lower frequencies and a dispersive region at higher frequencies. The plateau represents the frequency-independent region from which DC conductivity (σ_{dc}) is obtained, while the dispersion corresponds to bulk relaxation processes and the mid-frequency plateau corresponds to a frequency-independent conductivity region. As shown in Figure 3.5, σ_{dc} increases with lithium iodide concentration, with the 15wt.% LiI doped composition showing the highest room-temperature conductivity. At lower frequencies, conductivity decreases due to polarization effects at the electrodes and the restricted motion of ions near the electrode–electrolyte boundary, whereas the high-frequency domain exhibits enhanced conductivity due to faster ionic responses.

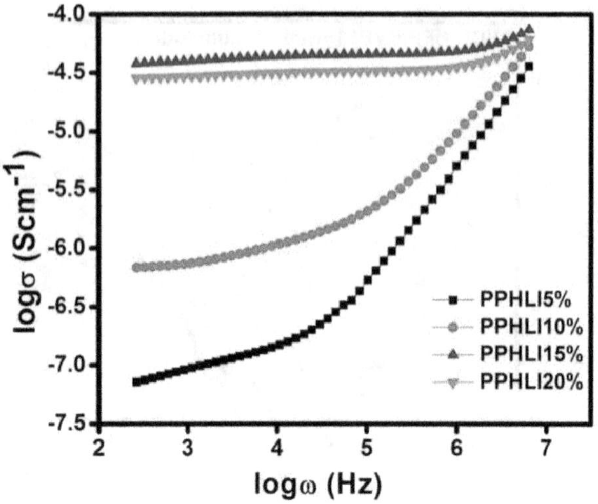

Figure 3.6 Conductance spectra for PPHLI5 wt.%, PPHLI10 wt.%, PPHLI15 wt.%, and PPHLI20 wt.% SPE.

Figure 3.7 further confirms that the conductivity of PPHLI 15 wt.% improves with temperature, primarily due to the plasticizing role of PVP, which enhances segmental motion, ionic mobility, and charge carrier density [11–13].

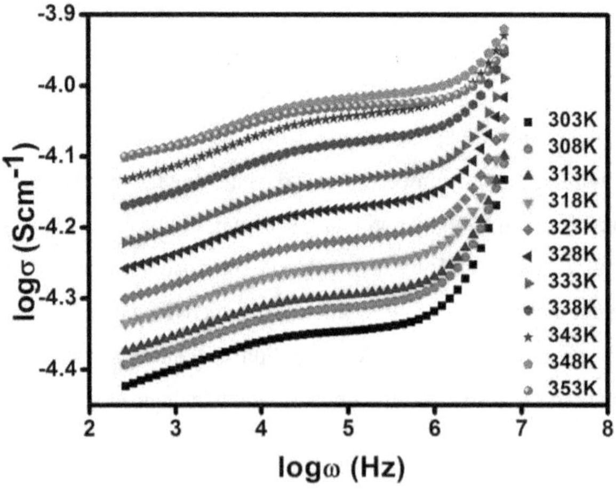

Figure 3.7 Conductance spectra of PPHLI15 wt.%, heat-treated at different temperatures.

3.3.3.3 Analysis of ionic conductivity as a function of temperature

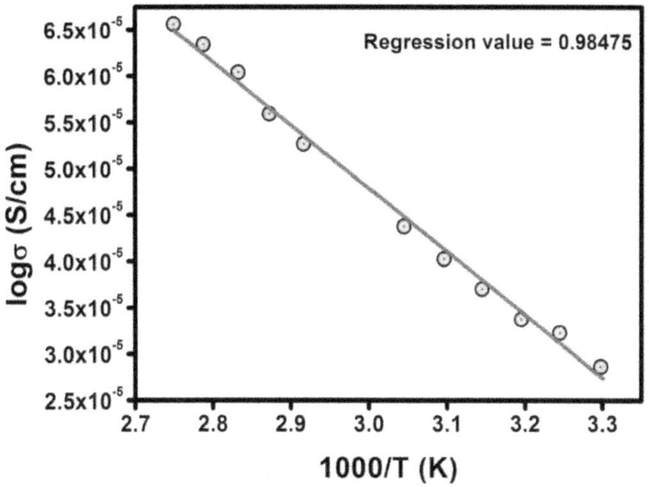

Figure 3.8 Representation of the Arrhenius behaviour for the PPHLI15 wt.% polymer blend electrolyte.

The synthesized polymer blend electrolytes were investigated with respect to their ionic conductivity across the temperature range of 303–343 K. Figure 3.8 illustrates the plot of log(σT) against 1000/T for all compositions, which demonstrates a nearly linear trend, confirming Arrhenius-type thermally activated conduction. This behaviour indicates that the mobility of charge carriers is strongly influenced by thermal energy, where increasing temperature assists in overcoming the potential barriers for ion hopping within the polymer matrix. The conduction behaviour indicates that ion transport is supported by polymer chain segmental motion, while elevated temperatures enhance chain flexibility and free volume, resulting in reduced activation energy for ionic migration.

Ea, the activation energy, is calculated by evaluating the slope from the equation

$$\sigma T = \sigma_0 \exp\left(-\frac{E_a}{kT}\right) \quad \text{............ 3.2}$$

where σ denotes the ionic conductivity, T is the absolute temperature, σ_0 corresponds to the pre-exponential factor, Ea indicates the activation energy associated with ion transport, and k represents Boltzmann's constant.

Taking logarithms gives a straight-line form versus reciprocal temperature:

$$\ln(\sigma T) = \ln \sigma_0 - \frac{E_a}{k}\left(\frac{1}{T}\right) \quad \text{......... 3.3}$$

or, in base-10,

$$\log_{10}(\sigma T) = \log_{10} \sigma_0 - \frac{E_a}{2.303\,k}\left(\frac{1}{T}\right) \quad \text{........ 3.4}$$

Thus, a plot of $\log_{10}(\sigma T)$ vs. 1000/T should be linear with slope m; the activation energy (E_a) can be calculated from:

$E_a = -$ mk (with m from ln plot)

or

$E_a = -$ m(2.303k) (with m from base-10 log plot)

For molar units,

E_a ($kJmol^{-1}K^{-1}$) = - m R x 10^{-3}, from ln(σT) vs 1/T

or

E_a ($kJmol^{-1}K^{-1}$) = -m (2.303R) x 10^{-3}, from $\log_{10}(\sigma T)$ vs. 1/T

with R=8.314 J $mol^{-1}K^{-1}$

The observed linearity indicates thermally activated hopping, increasing temperature enhances polymer-segment mobility and free volume, lowering barriers for Li^+ migration.

3.3.3.4 Dielectric Studies

i) Frequency-Dependent Behaviour of Dielectric Parameters

The dielectric constant (ε') and dielectric loss (ε'') are key parameters for understanding ion transport and polymer–salt interactions in solid polymer electrolytes. The dielectric constant is directly related to the ability of the polymer matrix to dissociate salts and stabilize free ions; a higher ε' typically reduces ion–ion electrostatic coupling and suppresses crystalline salt formation.

As shown in Figure 3.9(a), ε′ exhibits very high values in the low-frequency region and decreases progressively with increasing frequency until it reaches a nearly constant level at higher frequencies. This pronounced dielectric response at low frequencies is mainly due to space charge or interfacial polarization, where mobile ions accumulate at the electrode–electrolyte interfaces. At elevated frequencies, the swift alternation of the applied electric field restricts ions from tracking the field reversals, thereby diminishing electrode polarization and resulting in a decrease in ε′. Figure 3.9(b) further supports this observation by showing the corresponding behaviour of dielectric loss (ε″) across the frequency spectrum.

ii) Effect of Salt Concentration

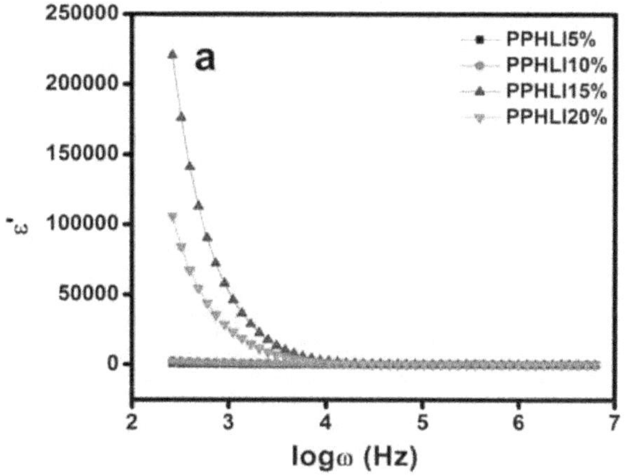

Figure 3.9 (a) Dielectric constant of PVDF-HFP: PVP blend polymer electrolytes containing different LiI compositions.

The dielectric response is also strongly influenced by salt content in the blend matrix. With increasing temperature, the polymer chains gain flexibility, reducing PVDF's crystalline domains and favouring amorphous regions that enhance ion transport. Nevertheless, when the LiI-salt concentration exceeds 15 wt.%, the dielectric constant in the low-frequency region decreases noticeably. This reduction indicates the formation of ion clusters or aggregates, which restrict effective ion dissociation and impede mobility. The occurrence of aggregation obstructs ion transport pathways, thereby suppressing ionic conductivity and adversely affecting the functional performance of the polymer electrolyte system.

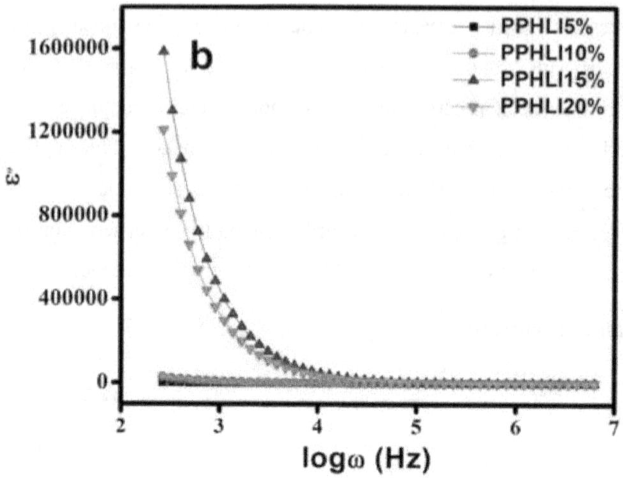

Figure 3.9 (b) Dielectric loss for various compositions of LiI mixed PVDF-HFP:PVP blend polymer electrolytes.

3.3.3.5 Real and Imaginary Modulus Studies

The study of electric modulus spectra provides valuable insights into the dynamics of ion transport within polymer electrolyte networks. Figure 3.10 (a, b) presents electric modulus spectra of PVDF-HFP/PVP/LiI-based solid polymer electrolytes at different salt concentrations. The real part of the modulus (M′) tends toward zero at low frequencies due to minimal electrode polarization [14,15]. While the imaginary component (M″) exhibits a distinct relaxation peak, which can be attributed to the coupling between polymer segmental motion and ion migration [16,17]. This relaxation peak is associated with long-range ion migration through the polymer matrix. With increasing salt concentration, the M″ peak gradually shifts to higher frequencies, signifying improved ion mobility and a relaxation process governed by thermal activation.

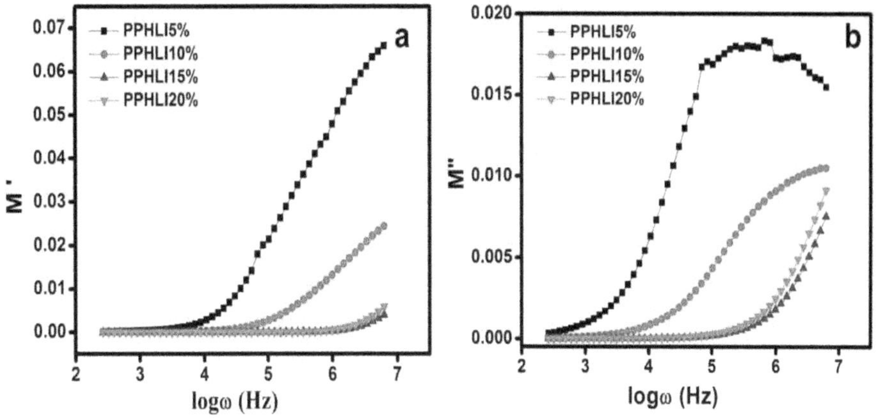

Figure 3.10 (a, b) Real and imaginary moduli of PPHLI5 wt.%, PPHLI10 wt.%, PPHLI15 wt.%, and PPHLI20 wt.% polymers.

3.3.3.6 Tangent Studies

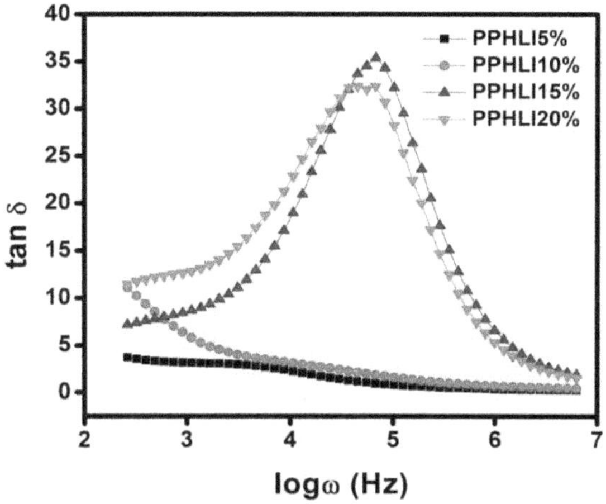

Figure 3.11. Tangent analysis for PPHLI5 wt.%, PPHLI10 wt.%, PPHLI15wt.%, and PPHLI20wt.%SPE.

The frequency-dependent behaviour of Tan δ for PVDF-HFP/PVP/LiI blend electrolytes at 303 K is presented in Figure 3.11. In all samples, the loss tangent exhibits an initial rise with frequency, followed by a pronounced maximum, after which it decreases with further frequency increase. A shorter relaxation (τ) time implies faster ion dynamics and, consequently, enhanced ionic conductivity.

Among the studied systems, the 15wt.% LiI doped SPE composition exhibited the shortest relaxation time, which is consistent with its highest observed ionic conductivity [11]. The broad dispersion noted in the low-frequency domain indicates electrode polarization effects. Meanwhile, the pronounced intensity of the Tan δ peak arises from the high probability of ionic hopping events per unit time, which facilitates efficient charge transport [18].

3.3.3.7 Argand Plot Studies

The Argand plots obtained at different temperatures provide insight into the relaxation dynamics of the polymer electrolyte system (Figure 3.12). The plots exhibit incomplete semicircular arcs, a clear indication of non-Debye type relaxation. Such behaviour arises from the coexistence of multiple relaxation mechanisms and diverse relaxation processes originating from dipole–ion and ion–ion interactions [19,20].

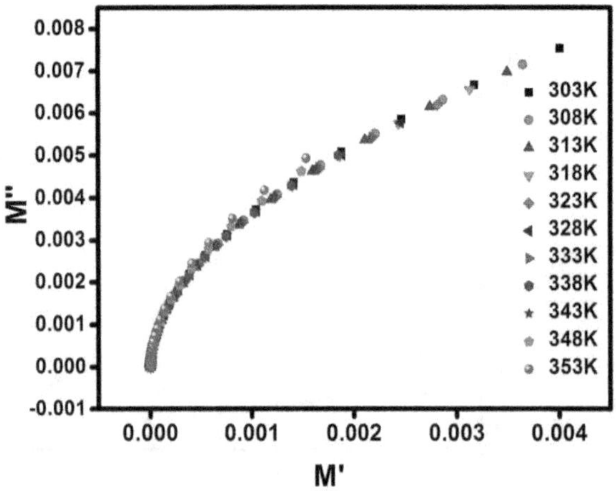

Figure 3.12 Argand plot for the PPHLI15 wt.% blend polymer electrolyte at various temperatures.

In the PVDF-HFP/PVP system, the dielectric response is strongly influenced by polymer polarity. The highly polar CF_2 groups of PVDF-HFP facilitate salt dissociation, while the carbonyl groups of PVP assist in dipolar alignment and Li^+ coordination. Together, these effects enhance dielectric permittivity and promote charge carrier generation, particularly at low frequencies where electrode polarization and space charge accumulation dominate.

The deviation from a perfect semicircle i.e. from ideal Debye relaxation reflects a broad distribution of relaxation times, consistent with the disordered and amorphous nature of the blended polymer. This indicates that ion transport occurs through segmental motion and hopping processes, which are thermally activated and heterogeneous in nature, thereby confirming the non-Debye relaxation behaviour of the system [15, 16].

Table 3.3 Over all AC impedance results

Sample	**Conductivity (S·cm−1)**	**Activation Energy**	**Dielectric Constant**	**Relaxation Time (S)**
PPHLI 5%	1.2385×10^{-7}	-2.843×10^{-9}	2.876×10^{3}	2.517×10^{-4}
PPHLI 10%	6.8678×10^{-7}	-7.529×10^{-7}	2.918×10^{4}	3.942×10^{-4}
PPHLI 15%	7.504×10^{-5}	-6.824×10^{-5}	1.586×10^{6}	1.740×10^{-4}
PPHLI 20%	2.837×10^{-5}	-8.195×10^{-6}	1.21173×10^{5}	1.440×10^{-4}

A reduction in arc length with increasing temperature confirms enhanced ionic conductivity due to thermally activated mobility. This correlation between the arc radius and conductivity highlights the strong dependence of ion mobility on thermal activation [21, 22]. Complementary analysis (Table 3.3) further shows that conductivity increases with LiI concentration up to 15 wt.%, accompanied by reduced activation energy, higher dielectric constant, and shorter relaxation time. These trends establish the 15wt.% LiI doped polymer blend composition as the most efficient electrolyte, exhibiting superior ionic transport properties.

3.4 Conclusions

A composite SPE was prepared via solution casting by integrating lithium iodide (LiI) into a PVDF-HFP/PVP blend for potential use in lithium-ion batteries. Extensive structural and electrical analyses were performed to evaluate the properties of the material. XRD analysis verified the complex structural arrangement of the host polymers and the incorporated salt, revealing that the introduction of LiI up to 15 wt.% enhanced the amorphous phase of the electrolyte matrix. FTIR spectroscopy confirmed the characteristic functional groups and revealed polymer–salt interactions through subtle peak shifts. Among the prepared compositions, the electrolyte containing 15 wt.% LiI (PPHLI15) exhibited the highest ionic conductivity, reaching **7.504×10^{-5} S·cm^{-1}**. The temperature dependence of conductivity followed an Arrhenius-type behaviour, and the applicability of the modified Arrhenius relation confirmed that ionic transport is strongly linked with the dielectric response of the polymer system. Dielectric studies further demonstrated that higher dielectric constants in the polymer host facilitate increased charge carrier density, thereby improving ionic conductivity. Owing to its favourable balance of ionic transport, electrochemical stability, and intermediate band gap, the PVDF-HFP/PVP–LiI (15 wt.%) electrolyte composition demonstrates strong potential as a viable solid electrolyte for next-generation lithium-ion battery systems.

3.5 References

[1] Abbrent, S.; Plestil, J.; Hlavata, D.; Lindgren, J.; Tegenfeldt, J.; Wendsjo, A. Crystallinity and morphology of PVdF-HEP-based gel electrolytes. *Polymers* **2001**, *42*, 1407–1416.

[2] Karpagavel, K.; Sundaramahalingam, K.; Manikandan, A.; Vanitha, D.; Manohar, A.; Nagarajan, E.R.; Nallamuthu, N. Electrical properties of lithium-ion conducting poly(vinylidene fluoride-co-hexafluoropropylene) (PVDF-HEP)/polyvinylpyrrolidone (PVP) solid polymer electrolyte. *J. Electron. Mater.* **2021**, *50*, 4415–4425.

[3] Ubarhande, R.M.; Bhattacharya, S.; Rani, M.U.; Babu, R.S.; Krishnaveni, S. Investigations on PVdF-HFP-PEMA polymer blend electrolytes doped with different lithium salts. *IOP Conf. Ser. Mater. Sci. Eng.* **2017**, *263*, 022004.

[4] Sangeetha, M.; Mallikarjun, A.; Jaipal Reddy, M.; Siva Kumar, J. FTIR spectroscopic and DC ionic conductivity studies of PVDF-HEP:LiBF4:EC plasticized polymer electrolyte membrane. *IOP Conf. Ser. Mater. Sci. Eng.* **2017**, *225*, 012049.

[5] Ulaganathan, M.; Rajendran, S. Effect of different salts on PVAc/PVdF-co-HEP based polymer blend electrolytes. *J. Appl. Polym. Sci.* **2010**, *118*, 646–651.

[6] Ramesh, S.; Yahaya, A.H.; Arof, A.K. Dielectric behaviour of PVC-based polymer electrolytes. *Solid State Ionics* **2002**, *152–153*, 291–294.

[7] Karger, J. Fast ion transport in solids. *Z. Phys. Chem.* **1995**, *189*, 274–275.

[8] Shyly, P.M.; Dawn Dharma Roy, S.; Thiravetyan, P.; Thanikaikarasan, S.; Sebastian, P.J.; Eapen, D.; Sahaya Shajan, X.J. Investigations on the effect of chitin nanofiber in PMMA-based solid polymer electrolyte systems. *New Mater. Electrochem. Syst.* **2014**, *17*, 133.

[9] Shalu, S.K.; Chaurasia, S.K.; Singh, R.K.; Chandra, S. Thermal stability, complexing behavior, and ionic transport of polymeric gel membranes based on polymer PVdF-HFP and ionic liquid [BMIM][BF4]. *J. Phys. Chem. B* **2013**, *117*, 897–906.

[10] Sim, L.N.; Majid, S.R.; Arof, A.K. FTIR studies on PEMA/PVdF-HEP blend polymer electrolyte system incorporated with LiCF3SO3 salt. *Vibr. Spectrosc.* **2012**, *58*, 57–66.

[11] Inbavalli, D.; Selvasekarapandian, S.; Sanjeeviraja, C.; Baskaran, R.; Nithya, S.; Kawamura, J.; Masuda, Y. Analysis of P(VdCl-coAN-co-MMA)-LiClO4-EC triblock copolymer electrolytes. *Bull. Mater. Sci.* **2015**, *38*, 183–190.

[12] Chen-Yang, Y.W.; Chen, H.C.; Lin, F.J.; Liao, C.W.; Chen, T.L. Preparation and conductivity of the composite polymer electrolytes based on poly[bis(methoxyethoxyethoxy)phosphazene], LiClO4 and α-Al2O3. *Solid State Ionics* **2003**, *156*, 383–392.

[13] Liao, C.S.; Ye, W.-B. Structural and conductive properties of poly(ethylene oxide)/layered double hydroxide nanocomposite polymer electrolytes. *Electrochim. Acta* **2004**, *49*, 4993–4998.

[14] Mahalakshmi, P.; Chitra, S.; Radha, K.P. Dielectric and ionic conductivity analysis of solid polymer electrolyte based on PMMA. *Int. J. Adv. Sci. Res.* **2016**, *1*, 21–24.

[15] Sangeetha, R.S.D.; Arasu, P.T.; Hirankumar, G.; Bella, R.S.D. Analysis of dielectric, modulus, electrochemical stability of PVP–ABSA polymer electrolyte systems. *Int. J. Chem. Sci.* **2016**, *14*, 477–481.

[16] Woo, H.J.; Majid, S.R.; Arof, A.K. Dielectric properties and morphology of polymer electrolyte based on poly(ε-caprolactone) and ammonium thiocyanate. *Mater. Chem. Phys.* **2012**, *134*, 755–761.

[17] Radha, K.P.; Selvasekarapandian, S.; Karthikeyan, S.; Hema, M.; Sanjeeviraj, C. Synthesis and impedance analysis of proton-conducting polymer electrolyte PVA:NH_4F. *Ionics* **2013**, *19*, 1437–1447.

[18] Sundaramahalingam, K.; Vanitha, D.; Nallamuthu, N.; Manikandan, A.; Muthuvinayagam, M. Electrical properties of lithium bromide polyethylene oxide/poly pyrrolidone polymer blend electrolyte. *Phys. B Condens. Matter* **2019**, *553*, 120–126.

[19] Aziz, S.B.; Abidin, Z.H.Z.; Arof, A.K. Influence of silver ion reduction on electrical modulus parameters of solid polymer electrolyte based on chitosan silver triflate electrolyte membrane. *Express Polym. Lett.* **2010**, *4*, 300–310.

[20] Pradhan, D.K.; Choudhary, R.N.P.; Samantaray, B.K. Studies of structural, thermal and electrical behavior of polymer nanocomposite electrolytes. *Express Polym. Lett.* **2008**, *2*, 630–638.

[21] Hill, R.M.; Dissado, L.A. Debye and non-Debye relaxation. *J. Phys. C Solid State Phys.* **1985**, *18*, 3829.

[22] Aziz, B.S.; Abdullah, O.G.; Hussein, S.A.; Ahmed, H.M. Effect of PVA blending on structural and ion transport properties of CS:AgNt-based polymer electrolyte membrane. *Polymers* **2017**, *9*, 622.

Chapter 4

Structural and Electrical Investigations of PVDF-HFP/PVP–$LiNO_3$ Solid Polymer Electrolytes

4.1 Introduction

A comprehensive study on PVDF-HFP/PVP-based SPEs doped by lithium iodide (LiI) was presented. The investigation highlighted the effectiveness of the PVDF-HFP/PVP matrix in promoting ionic conduction through its balanced combination of mechanical stability, dielectric strength, and amorphous content, while also demonstrating that the addition of LiI significantly enhanced salt dissociation and overall ionic conductivity. The results established that polymer–salt interactions and relaxation dynamics play a critical role in determining electrochemical performance.

Based on the insights gained from the earlier investigation, the present chapter expands the study by examining lithium nitrate ($LiNO_3$) as an alternative dopant within the same PVDF-HFP/PVP host framework. The choice of $LiNO_3$ is guided by its distinct physicochemical properties relative to lithium iodide (LiI). While lithium iodide (LiI) demonstrated effective salt dissociation and notable coordination with the polar moieties of the polymer blend, the relatively larger iodide anion and its distinct interaction characteristics impose constraints on achieving optimal ion transport.

In contrast, $LiNO_3$ possesses characteristics that make it particularly suitable for polymer electrolyte systems. Its high dielectric constant promotes effective salt dissociation, while its smaller, strongly solvating nitrate anion enables stronger interactions with the polymer backbone, thereby enhancing charge carrier availability. Furthermore, its chemical compatibility with polymer hosts ensures stable incorporation into the PVDF-HFP/PVP framework.

The introduction of $LiNO_3$ is therefore expected to influence multiple aspects of the electrolyte system. Structurally, it can disrupt crystallinity and increase amorphous content, thereby facilitating ion migration. Dielectrically, it may enhance polarization and modulate relaxation dynamics by strengthening polymer–ion coordination. Electrically, it is anticipated to improve ionic conductivity by providing more efficient conduction pathways within the amorphous regions. Additionally, $LiNO_3$ has been associated with improved thermal stability and favorable relaxation behavior, further supporting its selection for this study.Through this systematic evaluation, the chapter aims to elucidate how $LiNO_3$ incorporation alters the structural arrangement, dielectric response, and conductivity performance of PVDF-HFP/PVP polymer electrolytes, and to compare these findings to provide deeper insights into how the choice of lithium salt governs ion conduction mechanisms and influences the electrochemical performance of blended polymer electrolytes.

4.2 Preparation and Experimental Methods

4.2.1 Preparation of PVDF-HFP/PVP/$LiNO_3$ Solid Polymer Electrolytes

For the synthesis of the SPE membrane, a PVDF-HFP/PVP blend ratio of 60:40 wt.% was maintained as the host matrix. Lithium nitrate ($LiNO_3$) was incorporated at varying concentrations of 5, 10, 15, 20, and 25 wt.%, and the resulting samples were designated as LiN5, LiN10, LiN15, LiN20, and LiN25, respectively.

The electrolytes were fabricated using the directly solution casting method. Primarily, predetermined amounts of PVDF-HFP, PVP, and $LiNO_3$ were weighed accurately and dissolve separately in dimethylformamide (DMF) solution under continuous stirring for approximately three hours, ensuring complete dissolution. The individual solutions were then combined and further stirred for nearly six hours to promote homogeneity and induce gelation.

The resulting homogeneous solution was carefully poured into clean Petri dishes and allowed to undergo slow solvent evaporation under ambient conditions, leading to the formation of thin polymer electrolyte films. To remove residual solvent and enhance film stability, the samples were subsequently dried under vacuum. The obtained films were transparent, flexible, and uniform in surface morphology, making them suitable for further structural, dielectric, and electrical characterization. Figure 4.1 shows the preparation method of Blend polymer electrolyte film.

4.2.2 Characterization Techniques

The prepared PVDF-HFP/PVP/$LiNO_3$ solid polymer electrolyte films were subjected to a series of structural, spectroscopic, dielectric, and electrical characterizations to evaluate their physicochemical and electrochemical performance. The following procedure was employed:

- X-ray Diffraction (XRD)
- Fourier Transform Infrared Spectroscopy (FTIR)
- AC Impedance Spectroscopy
  - ➢ Conductivity studies
  - ➢ Conductance Spectra Analysis

- Dielectric Analysis
- Loss Tangent (tan δ) Analysis
- Argand Plot Analysis

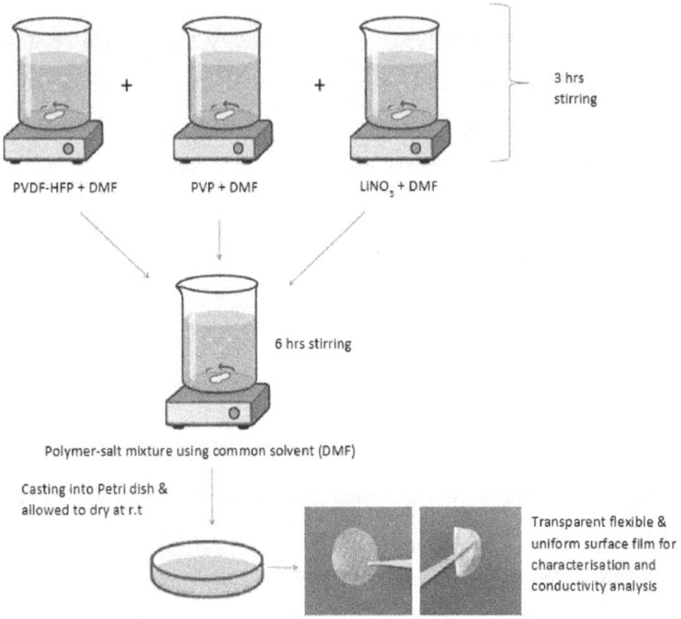

Figure 4.1 Solution casting method for preparation of PVDF-HFP/ PVP/$LiNO_3$solid polymer electrolytes.

4.3. Results and discussion

4.3.1 X-ray Diffraction (XRD) Analysis

The XRD patterns of the PVDF-HFP/PVP polymer blend incorporated with varying concentrations of $LiNO_3$, designated as LiN5, LiN10, LiN15, LiN20, and LiN25 is presented in Figure 4.2. The characteristic crystalline reflections of pure PVDF-HFP are observed at diffraction angles around 18°, 20°, and 27° [1]. Upon blending with PVP, notable modifications appear in the diffraction profile: the peak near 18° becomes more diffuse, the reflection at 20°

broadens, and additional broad humps emerge in the region of 30–40°, indicative of increased amorphous content in the blend matrix.

The presence of PVP, an amorphous polymer, acts as a disruptive agent to the crystalline domains of PVDF-HFP, thereby reducing the intensity and sharpness of the characteristic peaks. This effect is analogous to the role of plasticizers, where suppression of crystallinity enhances chain flexibility and improves ion mobility.

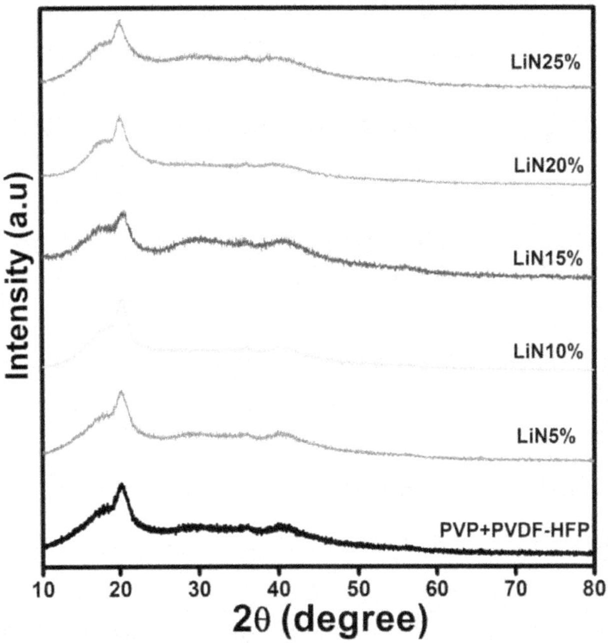

Figure 4.2 XRD patterns of PVDF-HFP (60 wt.%): PVP (40 wt.%) blends with different concentrations of $LiNO_3$ (5, 10, 15, 20 and 25 wt.%).

Upon the incorporation of lithium nitrate, further modifications are evident. Specifically, the intensity of the peak near 27° diminishes, while the broad hump at approximately 40° becomes more prominent. These observations suggest that $LiNO_3$ interacts with the host polymer chains, reducing ordered crystalline regions and promoting the formation of amorphous domains. The suppression of crystallinity, reflected by lower peak intensities, is strongly correlated with an increase in amorphous phase content.

Such structural changes are of significant importance, as amorphous regions provide favourable pathways for ionic conduction. The reduction in crystallinity, coupled with enhanced amorphous character, facilitates the dissociation of lithium salts and the migration of charge carriers. Thus, the XRD results clearly demonstrate that the integration of $LiNO_3$ into the PVDF-HFP/PVP matrix disrupts polymer chain packing, leading to structural modifications that underpin improved ionic transport properties [2].

4.3.2 Fourier Transform Infrared (FTIR) Spectroscopy

FTIR examination was performed to study the molecular relations and confirm the complexation behaviour within the PVDF-HFP/PVP/$LiNO_3$ solid polymer electrolyte system. Variations in peak position, shape, and intensity provide clear evidence of modifications in the polymer structure upon salt incorporation. The FTIR spectra of PVDF-HFP, PVP, and their $LiNO_3$-based complexes are illustrated in Figure 4.3.

The characteristic vibrational features of the host polymers were distinctly observed. For PVDF-HFP, the –CF_2 antisymmetric stretching mode appeared at 1168 cm^{-1}, while out-of-plane –CF_3 bending vibrations were identified at approximately 1072 cm^{-1}. In the case of PVP, the strong C=O stretching band was located around 1664 cm^{-1}, reflecting the polar nature of the pyrrolidone ring [3]. Additionally, weaker absorption bands in the range of 1384–1400 cm^{-1}, related to –CH_2 wagging and C–F stretching, were observed to gradually diminish in intensity with increasing salt concentration.

Further structural features include the bands at 1290 cm^{-1}, 830 cm^{-1}, and 737 cm^{-1}, which are attributed to the α-phase crystalline modes of PVDF-HFP [4]. The broad band detected at 3440 cm^{-1} is associated with O–H stretching vibrations, originating from PVP. Importantly, blending with varying concentrations of $LiNO_3$ led to clear shifts in peak position, intensity, and profile, confirming active polymer–salt interactions. Particular significance is the C=O stretching mode in the pyrrolidone ring region (1660–1650 cm^{-1}), which exhibited progressive shifts and intensity changes with increasing $LiNO_3$ content. The growth of this peak around 1660 cm^{-1} indicates strong coordination between Li^+ ions and the carbonyl groups of PVP, thereby validating effective salt–polymer complexation [5,6]. Furthermore, the slight shift of the vibrational band from 854 cm^{-1} to 858 cm^{-1} upon $LiNO_3$ incorporation provides additional evidence of the integration of salt into the polymer matrix and the modification of the local bonding environment [7].

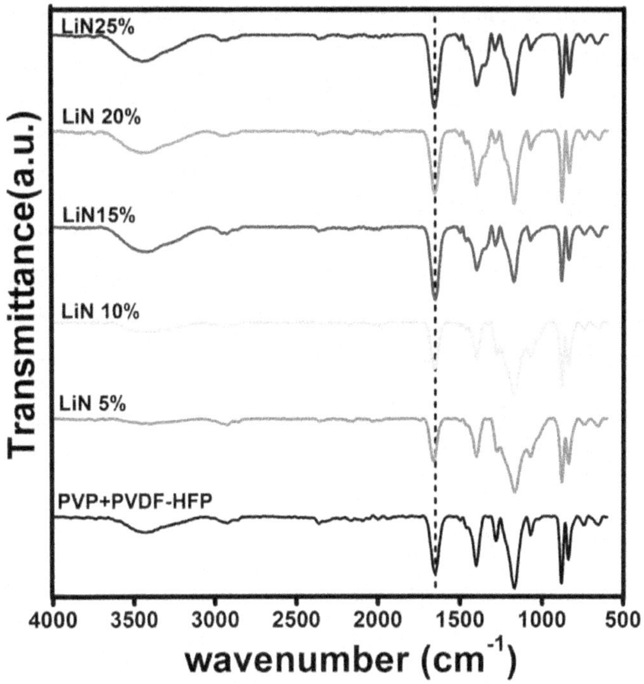

Figure 4.3 FTIR spectra of PVDF-HFP (60 wt.%): PVP (40 wt.%) blends with different concentrations of $LiNO_3$ (5, 10, 15, 20 and 25 wt.%).

4.3.3 Impedance Spectroscopy

4.3.3.1 Conductivity studies:

The Nyquist plots of the PVDF-HFP/PVP/$LiNO_3$ solid polymer electrolytes with compositions PVDF-HFP (60 wt.%): PVP (40 wt.%) with various concentrations of $LiNO_3$ (5, 10, 15, 20 and 25 wt.%) at ambient temperature are shown in Figure 4.4. All spectra exhibit a half semicircular arc in the high-frequency zone followed with a spike in the low-frequency zone, which is characteristic of ion-conducting polymer systems. The semicircular feature corresponds to the bulk response of the electrolyte, arising from the similar effect of resistance and capacitance elements, while the low-frequency spike reflects capacitive polarization effects at the electrode–electrolyte edge, typically associated with blocking electrodes.

The intercept of the semicircle with the real axis (Z′) provides the bulk resistance (R_b) of the electrolyte. By fitting the spectra using an equivalent circuit model through Z-View software, precise values of bulk resistance were extracted. The ionic conductivity was calculated from R_b values using equation

$$\sigma = l/R_bA \qquad \text{......... 4.1}$$

where 'l' is the thickness of the SPE, 'R_b' is the bulk resistance found from the x-axis point in the complex impedance plot of the AC circuit, and 'A' is the surface area of the SPE.

The results clearly indicate that ionic conductivity increases with rising $LiNO_3$ concentration, due to enhanced salt dissociation and increased free charge carrier density. Among the investigated compositions, the polymeric electrolyte blend with composition PVDF-HFP (60 wt.%): PVP (40 wt.%): $LiNO_3$ (20wt.%) exhibited the lowest bulk resistance and therefore the maximum ionic conductivity at RT.

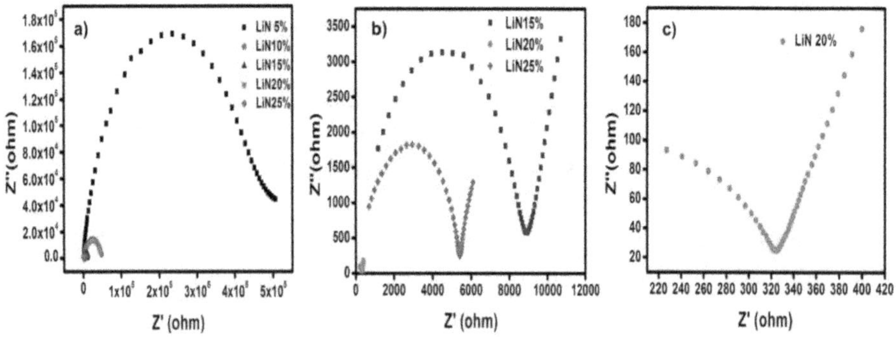

Figure 4.4 (a, b, c) Nyquist plots obtained by ac impedance for different concentration of $LiNO_3$ (5, 10, 15, 20 and 25 wt.%) doped PVDF-HFP/PVP SPEs at room temperature.

Temperature-dependent impedance analysis (Figure 4.5) further supports these observations. For the composition PVDF-HFP (60 wt.%): PVP (40 wt.%): $LiNO_3$ (20 wt.%), the diameter of the semicircle progressively reduces as the temperature rises from room temperature to 75 °C, signifying reduced resistance and enhanced ionic mobility. At 95 °C, the semicircle diameter increases slightly, indicating a temporary reduction in conduction efficiency, possibly due to ion clustering or restricted chain flexibility. Beyond 100 °C, the semicircle diameter decreases again, suggesting that additional thermal activation restores ionic mobility and facilitates more efficient ion transport.

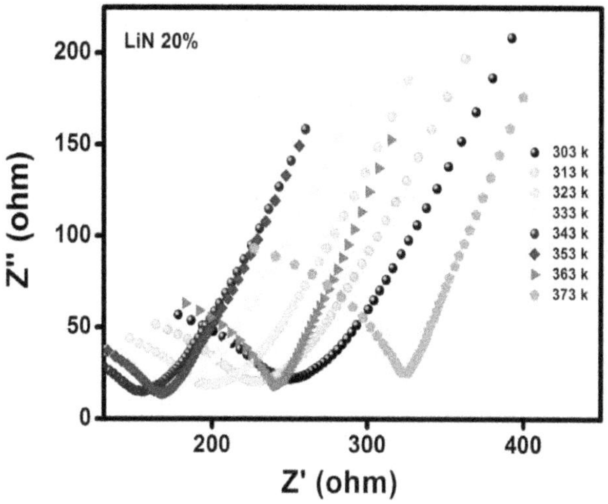

Figure 4.5 Impedance spectra for PVDF-HFP (60 wt.%): PVP (40 wt.%): $LiNO_3$ (20 wt.%), heat-treated at different temperatures.

The prepared $LiNO_3$ doped PVDF-HFP/PVP polymeric electrolyte blend with composition PVDF-HFP (60 wt.%): PVP (40 wt.%): $LiNO_3$ (20 wt.%) had a better high conductivity of 1.06×10^{-4} $S.cm^{-1}$.

The observed trends confirm that the incorporation of $LiNO_3$ not only reduces bulk resistance but also enhances temperature-activated ion dynamics within the polymer host matrix. This behaviour highlights the synergistic role of polymer–salt interactions and thermal energy in governing the ionic conduction mechanism of PVDF-HFP/PVP/$LiNO_3$ solid polymer electrolytes.

4.3.3.2 Temperature-Dependent Ionic Conductivity

To grow deeper near into the ionic conduction device of the PVDF-HFP/PVP/$LiNO_3$ polymer electrolyte, the temperature dependence of ionic conductivity was examined. Figure 4.6 presents the difference of log(σT) as a function of 1000/T for the LiN20 composition, calculated in the temperature range of 30–100 °C. The conductivity exhibits a non-linear trend with temperature. Initially, conductivity increases steadily up to 75 °C, followed by a reduction at 95 °C, and then a subsequent increase beyond 100 °C. At lower temperatures, the enhanced conductivity can be attributed to the combined contribution of Li^+ ions and residual H^+ ions, which improve charge carrier density. The decrease observed near 95 °C is likely due to the thermal decomposition of physiosorbed water molecules, important to a decrease in the number of mobile carriers. Above 100 °C, the renewed increase in conductivity is primarily connected with the enhanced thermal mobility of Li^+ ions, which dominate the transport process in the polymer matrix [8,9].

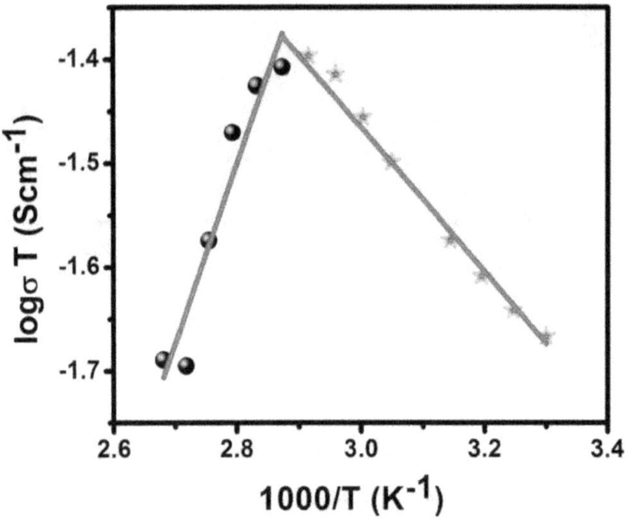

Figure 4.6 Arrhenius relationship for the LiN20wt% blend polymer electrolyte.

From the linear fitting of the Arrhenius plots, two distinct activation energy values were obtained: E_{a1} = 0.32663 eV and E_{a2} = 0.131765 eV, corresponding to different conduction regimes. The higher activation energy at lower temperatures is consistent with restricted ion motion, whereas the reduced activation energy at elevated temperatures reflects the facilitation of ion hopping within the amorphous regions of the polymer blend. These results confirm that the temperature-dependent conductivity of the prepared electrolyte follows Arrhenius behaviour, with conduction governed predominantly by thermally activated hopping of Li^+ ions [10].

4.3.3.3 Measurements of Conductance Spectra

The frequency-dependent conductivity spectra of PVDF-HFP/PVP polymer electrolytes doped with varying concentrations of $LiNO_3$ are presented in Figure 4.7. The spectra exhibit two distinct regions across all compositions: a dispersion zone at higher frequencies and a plateau region at lower frequencies. The intermediate frequency range represents the bulk relaxation phenomenon, which bridges the low-frequency plateau and the high-frequency dispersion domains. The dc conductivity (σ_{dc}) was extracted by extrapolating the low-frequency plateau region of the curve towards the y-axis.

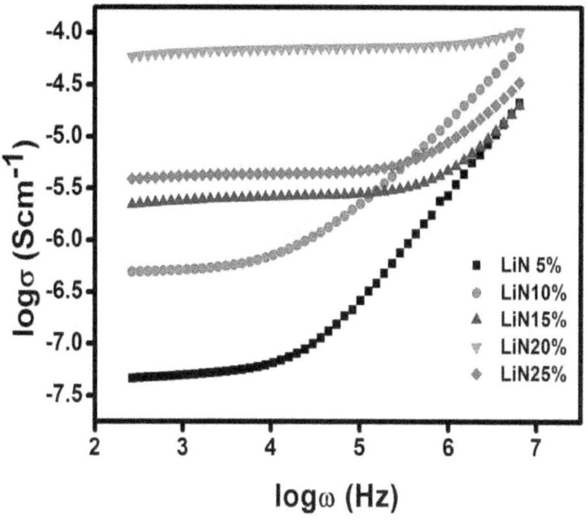

Figure 4.7 Conductance spectra for a PVDF-HFP/PVP different concentration of $LiNO_3$ salt.

As shown in Figure 4.7, the σ_{dc} values increase systematically with rising $LiNO_3$ concentration, confirming that higher salt content enhances ionic charge carrier density and mobility [11]. Among the investigated samples, the LiN20 composition displayed the highest room-temperature ionic conductivity, thereby establishing it as the most efficient electrolyte formulation within this study. Further insights into the conduction behaviour were obtained by examining the temperature dependence of the LiN20 sample, as depicted in Figure 4.8. The conductivity increases steadily with temperature up to 75 °C, after which a decline is observed at 95°C. Beyond this point, conductivity rises again at temperatures exceeding 100 °C. This non-linear behaviour may be interpreted as follows:

At lower temperatures, both Li^+ and residual H^+ ions contribute to conduction, enhancing ionic transport. The decline in conductivity near 95 °C can be credited to the thermal decomposition of adsorbed water molecules, which reduces the number of mobile carriers. At higher temperatures (>100 °C), the renewed rise in conductivity is linked to the increased thermal activation and mobility of Li^+ ions, which dominate charge transport in these polymer electrolytes [8,9].

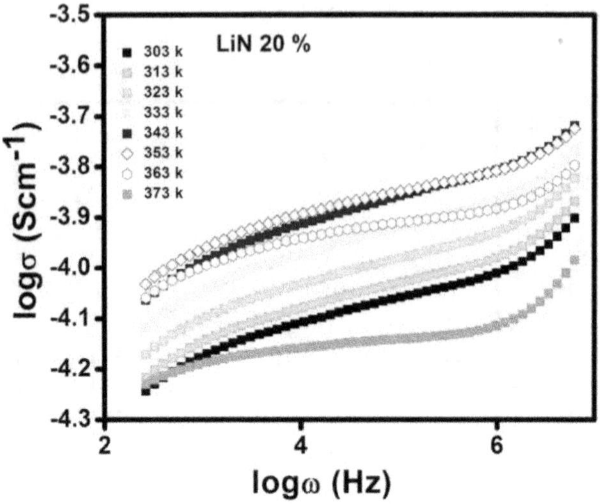

Figure 4.8 Conductivity spectra obtained for LiN20 wt% polymer blend electrolyte at different temperatures.

It is generally accepted that in such solid polymer electrolyte systems, Li^+ ions act as the primary charge carriers due to their strong coordination with electron-donating groups in the host matrix. The activation energy obtained from the dc conductivity plots suggests that surplus thermal energy facilitates additional Li^+ hopping, thereby improving the overall conduction mechanism [12].

4.3.3.4. Dielectric Studies

In polymer electrolytes, dielectric investigations provide a crucial link between structural modifications and charge transport behaviour. Dielectric loss (ε″) reflects the energy dissipated during ion migration when the applied field undergoes a rapid polarity reversal, whereas the dielectric constant (ε′) quantifies the ability of the system to store charge under an applied field [13]. The overall dielectric response is expressed through the complex permittivity (ε*), represented as:

$$\varepsilon^*(\omega) = \varepsilon'(\omega) - i\varepsilon''(\omega) \qquad \text{............ 4.2}$$

Here, ε′(ω) denotes the energy storage capability and ε″(ω) represents dielectric energy loss in each cycle of the applied field [14].

For the PVDF-HFP/PVP/$LiNO_3$ electrolyte system, the frequency-dependent performance of ε' and ε'' are demonstrate in Figure 4.9(a) & (b) in that order. At low frequencies, ε' display significantly elevated values, which gradually decline with rising frequency before stable at higher frequencies. This trend can be attributed to space-charge polarization, wherein mobile ions accumulate at the electrode–electrolyte interface, contributing to large permittivity values [15]. As the frequency increases, the alternating field prevents effective ion accumulation, resulting in a reduction of electrode polarization effects and a consequent drop in ε' [16].

A comparative analysis with the LiI-based system discussed in the preceding chapter reveals both similarities and distinctions. While both salts exhibit high ε' values at low frequencies due to interfacial polarization, the $LiNO_3$ system demonstrates a sharper decline in ε' with frequency. This behaviour suggests that the smaller nitrate anion disrupts ion–ion correlations more effectively than iodide, thereby enhancing salt dissociation and improving dielectric relaxation.

Temperature also plays an essential role in dielectric behaviour. As the temperature increases, polymer chain flexibility improves, reducing PVDF's crystalline domains and promoting ion transport. However, beyond the LiN20 composition, the dielectric constant at low frequencies decreases, which is attributed to the configuration of ion combined. Such cluster structure limits the availability of free ions and constrains the creation of effective conduction pathways, ultimately reducing charge transport efficiency.

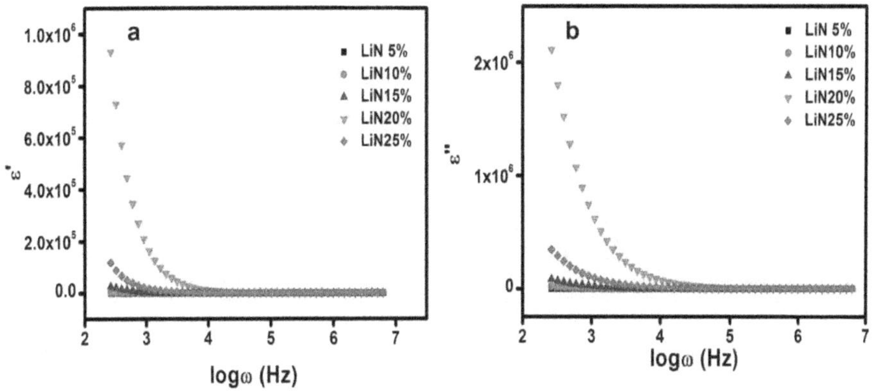

Figure 4.9 (a) & (b). Dielectric constant (ε') and dielectric loss (ε'') for different concentration of $LiNO_3$ (5, 10, 15, 20 and 25 wt.%) doped PVDF-HFP/PVP SPEs.

Thus, the dielectric response of the PVDF-HFP/PVP/$LiNO_3$ system demonstrates that $LiNO_3$ not only facilitates enhanced salt dissociation compared to LiI but also exhibits stronger frequency- and composition-dependent polarization effects. These result shows up the crucial position of anion size and interaction characteristics in governing dielectric behaviour and overall ionic conductivity in polymer electrolytes.

The dielectric properties of the PVDF-HFP/PVP polymer electrolyte containing 20 wt.% $LiNO_3$ are presented in Figure 4.10(a) and (b). Both the real part (ε′) and the imaginary part (ε″) of the permittivity exhibit a distinct increase in the low-frequency region as the temperature rises [17]. This enhancement reflects the combined influence of thermal activation, charge injection, and local polarization processes, all of which contribute to an increased density of movable charge transporter inside the method.

For the LiN20 composition, the frequency-dependent dielectric constant confirms that the charge accumulation near the electrode interface is effectively enhanced at elevated temperatures, consistent with the trend observed in the dc conductivity data. Notably, the dielectric constant exhibits minimal time dependence, indicating that the polarization processes are primarily governed by thermally activated ion migration rather than transient trapping or delayed relaxation effects.

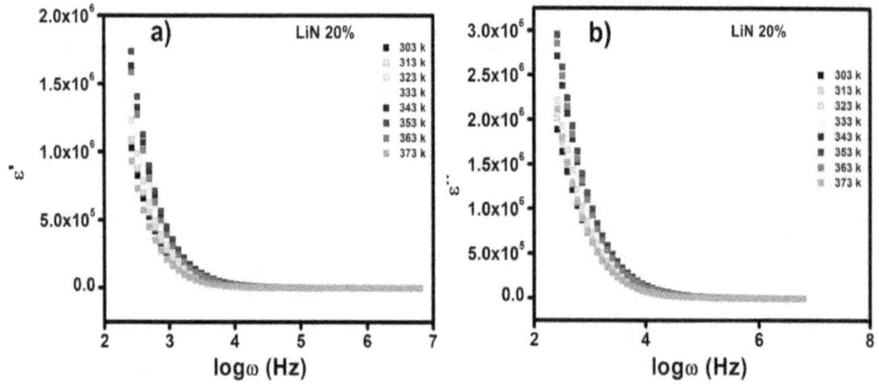

Figure 4.10 (a) &(b) Dielectric constant (ε′) and dielectric loss (ε″) for the LiN20 wt% blend polymer electrolyte at various temperatures.

A comparison with the LiI-based system discussed in the previous chapter reveals that, while both electrolytes display enhanced dielectric response with temperature, the nitrate-based system demonstrates a more pronounced ε′ and ε″ increase, signifying more effective salt dissociation and stronger electrode polarization effects. This difference may be accredited to the lesser size and stronger solvating ability of the nitrate anion, which facilitates greater Li^+ mobility within the amorphous polymer domains.

Thus, the dielectric analysis of the LiN20 composition highlights the synergistic role of salt concentration and temperature in enhancing ion dynamics, while also underlining the distinct advantages of $LiNO_3$ over LiI in modulating the dielectric reaction of the SPE.

4.3.3.5. Loss Tangent (tan δ) Analysis

The difference of the dielectric loss tangent (tan δ) by frequency for the PVDF-HFP/PVP electrolyte containing 20 wt.% $LiNO_3$ is illustrated in Figure 4.11. As the frequency increases, tan δ initially rises, reaching a maximum near 5 Hz, after which it decreases gradually to a minimum value. This peak behaviour is indicative of a relaxation process, wherein the frequency of the apply electric field approaches the natural relaxation frequency of movable charge transporter within the polymer matrix.

The influence of temperature on tan δ reveals a non-monotonic trend. Between 30 °C and 70 °C, tan δ decreases progressively, suggesting enhanced alignment and relaxation of charge carriers, possibly facilitated by both Li^+ and residual H^+ ions. Beyond 70 °C to 90 °C, tanδ increases again, which can be accredited to the thermal decomposition of adsorbed water molecules, introducing additional localized polarization and energy loss mechanisms. At temperatures exceeding 90 °C, the value of tan δ decreases once more. This decline is consistent with the enhanced mobility of Li^+ ions, which dominate charge transport at elevated temperatures and reduce polarization losses [8,9].

The observed behaviour signifies that the dielectric loss tangent in the LiN20 system is governed by a complex interplay of ion mobility, electrode polarization, and temperature-driven relaxation processes.

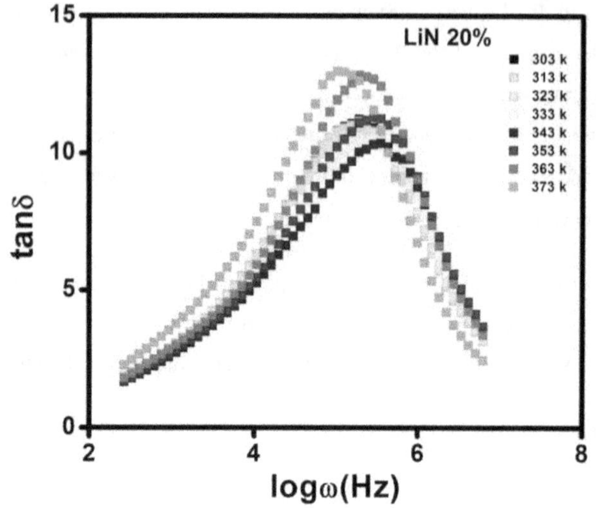

Figure 4.11 Tangent analysis for the LiN20 wt% blend polymer electrolyte at various temperatures.

4.3.3.6. Argand Plot Studies

To elucidate the nature of relaxation processes in the polymer electrolyte system, temperature-dependent Argand plots were examined for the PVDF-HFP/PVP blend containing 20 wt.% $LiNO_3$. The corresponding plots are presented in Figure 4.12. The observed arcs display incomplete semicircles, which deviate significantly from the idealized Debye model that assumes a single relaxation time. This deviation confirms the presence of non-Debye type relaxation, implying a division of relaxation times arising from the combined effects of ion–dipole interactions, polymer segmental dynamics, and electrode polarization phenomena.

The incomplete semicircular response suggests that polarization processes are not uniform across the system but are instead governed by multiple overlapping mechanisms. In particular, the strong interaction between mobile Li^+ ions and polar groups within the host matrix (CF_2 dipoles of PVDF-HFP and C=O groups of PVP) contributes to the observed broadening of the arcs. Such interactions enhance localized polarization, producing a complex dielectric relaxation landscape characteristic of amorphous and disordered polymer systems.

Furthermore, the length of the semicircular arc provides direct information about the extent of ionic transport. A shorter arc radius is associated with higher ionic mobility and lower bulk resistance, thereby signifying improved ionic conductivity. In the LiN20 composition, the reduced arc length at elevated temperatures clearly demonstrates the facilitated ion transport and the strong coupling between polymer flexibility and ionic conduction.

In summary, the Argand plot analysis confirms that the relaxation in PVDF-HFP/PVP/$LiNO_3$ electrolytes follows a non-Debye mechanism, reflecting the coexistence of multiple ionic environments and dynamic coordination states. The strong association between the arc length and conductivity further validates that the enhanced ionic transport originates from both structural modifications of the polymer matrix and effective dissociation of $LiNO_3$ within the host blend.

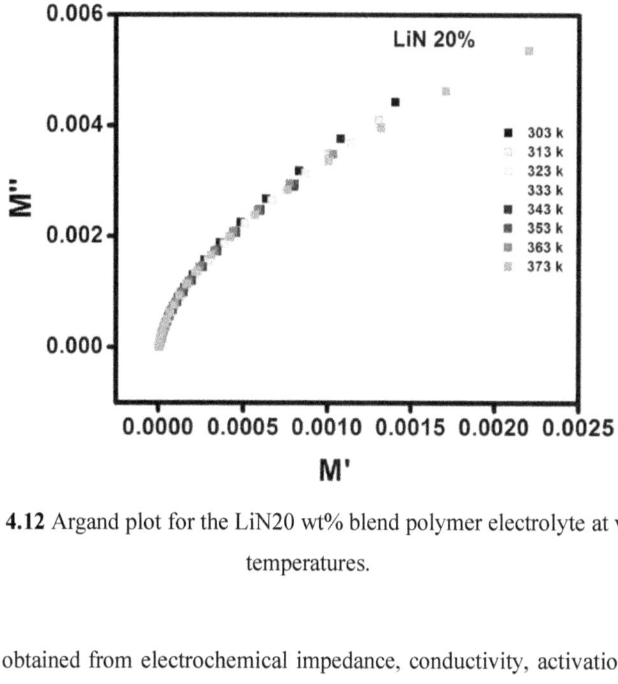

Figure 4.12 Argand plot for the LiN20 wt% blend polymer electrolyte at various temperatures.

The findings obtained from electrochemical impedance, conductivity, activation energy and dielectric studies, a comparative summary of the key parameters for the prepared $LiNO_3$-doped PVDF-HFP/PVP electrolytes is presented in Table 4.1.

Table 4.1 Over all AC impedance results

Sample	Conductivity ($S \cdot cm^{-1}$)	Activation Energy(eV)		Dielectric Constant
		E_{a1}	E_{a2}	
LiN 5%	8.02×10^{-8}	0.3266	0.1317	1.7751×10^{3}
LiN 10%	8.21×10^{-7}	0.2981	0.1252	2.0421×10^{3}
LiN 15%	4.38×10^{-6}	0.2172	0.0918	2.6314×10^{4}
LiN 20%	1.06×10^{-4}	0.1935	0.0821	9.3911×10^{5}
LiN 25%	8.96×10^{-6}	0.2083	0.0936	1.2412×10^{4}

4.4 Conclusion

SPEs based on PVDF-HFP/PVP with varying concentrations of lithium nitrate were successfully fabricated using the solution casting technique. Structural characterization confirmed the formation of a blended matrix, where XRD analysis revealed a reduction in crystallinity with increasing salt content, and FTIR spectra verified effective complexation between the host polymers and $LiNO_3$. Electrical investigations demonstrated that the incorporation of lithium nitrate significantly enhanced ion transport. The optimized composition exhibited a highest ionic conductivity of 1.06×10^{-4} $S \cdot cm^{-1}$, accompanied by activation energies of 0.32663 eV and 0.131765 eV, highlighting the thermally activated conduction mechanism. Dielectric and impedance analyses further established that the system follows a non-Debye relaxation process, governed by both polymer segmental motion and ion–dipole interactions. The combined results from impedance, conductance, and dielectric loss tangent studies confirmed the presence of efficient ionic relaxation dynamics within the polymer matrix, making PVDF-HFP/PVP–$LiNO_3$ electrolyte system a capable applicant for application in next-generation solid-state lithium-ion batteries.

4.5 References

[1] Erukhimovich, I.; de la Cruz, M.O. Phase equilibria and charge fractionation in polydisperse polyelectrolyte solutions. *J. Polym. Sci. B Polym. Phys.* **2004**, *42*, 2173–2187.

[2] Prabakaran, P.; Manimuthu, R.P.; Gurusamy, S.; Sebasthiyan, E. Plasticized polymer electrolyte membranes based on PEO/PVdF-HFP for use as an effective electrolyte in lithium-ion batteries. *Chin. J. Polym. Sci.* **2017**, *35*, 407–421.

[3] Mishra, K.; Hashmi, S.A.; Rai, D.K. Protic ionic liquid-based gel polymer electrolyte: Structural and ion transport studies and its application in proton battery. *J. Solid State Electrochem.* **2014**, *18*, 2255–2266.

[4] Gayen, A.L.; Improvisation of electrical properties of PVDF-HFP: Use of novel metallic nanoparticles. *J. Mater. Sci. Mater. Electron.* **2017**, *28*, 14798–14808.

[5] Kesavan, K.; Rajendran, S.; Mathew, C. M. Studies on poly (vinyl pyrrolidone) based solid polymer blend electrolytes complexed with various lithium salts. *Polymer Science Series B* **2014**, *56*, 520-529.

[6] Anandha Jothi, M.; Vanitha, D.; Nallamuthu, N.; Manikandan, A.; Asath Bahadur, S. Investigations of lithium ion conducting polymer blend electrolytes using biodegradable cornstarch and PVP. *Phys. B Condens. Matter* **2020**, *580*, 411940.

[7] Fuzlin, A. F.; Bakri, N. A.; Sahraoui, B.; Samsudin, A. S. Study on the effect of lithium nitrate in ionic conduction properties-based alginate biopolymer electrolytes. *Materials Research Express* **2019**, *7*, 015902.

[8] Wang, Q.; Ping, P.; Zhao, X.; Chu, G.; Sun, J.; Chen, C. Thermal runaway caused fire and explosion of lithium-ion battery. *J. Power Sources* **2012**, *208*, 210–224.

[9] Nallamuthu, N.; Prakash, I.; Venkateswarlu, M.; Balasubramanyam, S.; Satyanarayana, N. Sol-gel synthesis and characterization of Li2O-As2O5-SiO2 glassy system. *Mater. Chem. Phys.* **2008**, *111*, 24–28.

[10] Ahmad, N.H.B.; Isa, M.I.N.B.M. Proton conducting solid polymer electrolytes based on carboxymethyl cellulose doped ammonium chloride: Ionic conductivity and transport studies. *Int. J. Plast. Technol.* **2015**, *19*, 47–55.

[11] Masoud, E.M.; El-Bellihi, A.A.; Bayoumy, W.A.; Mousa, M.A. Organic-inorganic composite polymer electrolyte based on PEO-LiClO4 and nano-Al2O3 filler for lithium polymer batteries: Dielectric and transport properties. *J. Alloys Compd.* **2013**, *575*, 223–228.

[12] Boora, M. Investigation of structural and impedance spectroscopic properties of borate glasses with high Li+ concentration. *Solid State Ionics* **2021**, *368*, 115704.

[13] Prateek, V.K.; Thakur, R.K.; Gupta, R.K. Recent progress on ferroelectric polymer-based nanocomposites for high energy density capacitors: Synthesis, dielectric properties, and future aspects. *Chem. Rev.* **2016**, *116*, 4260–4317.

[14] Olhoeft, G.R. Electrical, magnetic, and geometric properties that determine ground penetrating radar performance. *Seventh Int. Conf. Ground Penetrating Radar* **1998**, 177–182.

[15] Li, C.; Gu, L.; Guo, X.; Samuelis, D.; Tang, K.; Maier, J. Charge carrier accumulation in lithium fluoride thin films due to Li-ion absorption by titania (100) subsurface. *Nano Lett.* **2012**, *12*, 1241–1246.

[16] Bhargav, P.B.; Sarada, B.A.; Sharma, A.K.; Rao, V.V.R.N. Electrical conduction and dielectric relaxation phenomena of PVA-based polymer electrolyte films. *J. Macromol. Sci. Part A Pure Appl. Chem.* **2010**, *47*, 131–137.

[17] Zannou, A.L.Temperature increases by kilohertz frequency spinal cord stimulation. *Brain Stimul.* **2019**, *12*, 62–72.

Chapter 5

Electrochemical Impedance Studies on PVA-PVP-Na_2SO_4 Solid Polymer Electrolytes

5.1 Introduction

In the preceding chapters, detailed investigations on PVDF-HFP/PVP-based polymer electrolytes doped with lithium salts (LiI and $LiNO_3$) established the strong potential of polymer blends in facilitating ionic transport through the formation of amorphous domains, effective polymer–salt complexation, and non-Debye relaxation behaviour [1-3]. While lithium-ion batteries (LIBs) remain the dominant technology in advanced electrochemical energy storage, issues related to the high cost of lithium, limited resource availability, and environmental concerns necessitate the exploration of alternative, more sustainable chemistries.

In this context, sodium-ion batteries (SIBs) have emerged as promising next-generation candidates due to the abundance, non-toxicity, low cost, and environmental friendliness of sodium [4]. Moreover, sodium shares similar electrochemical properties with lithium, making SIBs a practical and scalable alternative for large-scale energy storage applications. The challenge, however, lies in identifying suitable solid polymer electrolytes (SPEs) capable of supporting efficient sodium-ion transport with good electrochemical and mechanical stability.

To address this, the present chapter shifts focus from fluoropolymer-based systems to poly(vinyl alcohol) (PVA) and poly(vinyl pyrrolidone) (PVP) as the host matrix for solid polymer electrolytes. PVA is a semicrystalline polymer known for its chemical resistance, hydrogen bonding through –OH groups, and film-forming ability. However, its high crystallinity restricts ionic conduction at ambient temperature. PVP, in contrast, is an amorphous polymer containing polar carbonyl (C=O) groups, which promote ionic coordination and facilitate polymer chain flexibility. The PVA/PVP blend thus represents an attractive combination, wherein PVP reduces the crystallinity of PVA and provides additional dipolar sites for sodium-ion coordination.

To enable sodium-ion transport within the polymer matrix, sodium sulfate (Na_2SO_4) was selected as the dopant salt due to its efficient dissociation in polar polymer hosts, chemical stability, and environmentally friendly character [5]. Different concentrations of Na_2SO_4 were systematically incorporated into the PVA/PVP blend through the solution casting technique, resulting in the formation of flexible, free-standing solid polymer electrolyte films suitable for electrochemical characterization.

The present chapter focuses on the investigation of the conductivity behaviour of Na_2SO_4-doped PVA/PVP electrolytes, with particular emphasis on the role of salt concentration in governing impedance response, dielectric characteristics, and overall ionic transport properties. The outcomes of this study not only extend the insights gained from the lithium-based polymer systems discussed in earlier chapters but also broaden the scope of polymer blend electrolytes towards environmentally friendly sodium-ion battery applications.

5.2 Preparation of PVA/PVP–Na_2SO_4 Polymer Electrolyte Films

Solid polymer electrolyte membranes based on PVA/PVP in the ratio 80:20 blends doped with different concentrations (2, 4, 6, & 8 wt%) of sodium sulfate (Na_2SO_4) were synthesized using the solution casting technique, which is a simple and widely adopted method for fabricating uniform polymer films. High-purity polyvinyl alcohol (PVA), polyvinyl pyrrolidone (PVP), and sodium sulfate were employed as precursors.

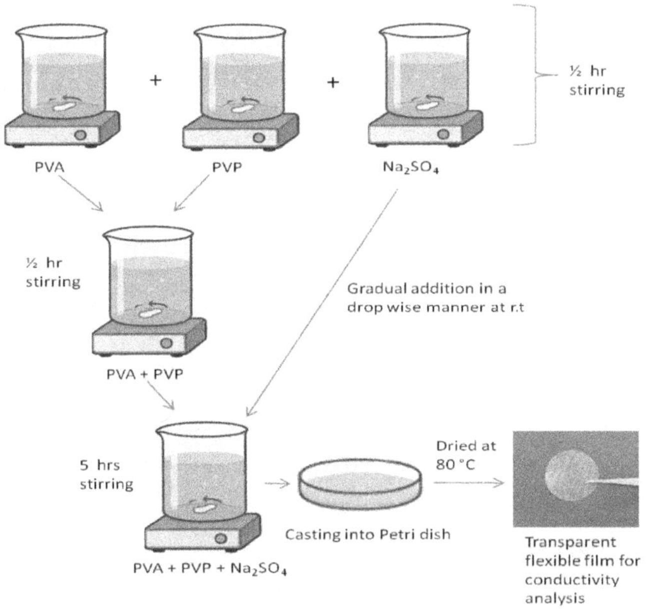

Figure 5.1 Solution casting method for preparation of PVA-PVP- Na_2SO_4 SPEs.

In the first step, required amounts of PVA, PVP, and Na_2SO_4 for the preparation of PVA-PVP-Na_2SO_4 SPEs were dissolved individually in double-distilled water under continuous magnetic stirring to ensure complete dissolution of each component. The aqueous solutions of PVA and

PVP were then combined and stirred for about 30 minutes to promote homogeneous blending of the two polymers. Following this, the Na_2SO_4 solution was added gradually in a dropwise manner at room temperature, allowing controlled incorporation of the salt into the polymer matrix.

The resulting mixture was stirred further for approximately five hours until a stable, viscous gel was obtained. This gel was cast onto clean glass Petri dishes and subjected to drying at 80 °C, leading to the slow evaporation of water and the formation of transparent, flexible polymer electrolyte films. After drying, these films were carefully peeled off and stored in airtight containers to prevent moisture absorption prior to further characterization. Figure 5.1 shows the preparation method of PVA-PVP- Na_2SO_4 solid polymer electrolyte films.

5.3 Results and discussion

5.3.1 X-ray Diffraction (XRD) Analysis

The X-ray diffraction (XRD) patterns of the PVA:PVP polymer blend with increase composition of Na_2SO_4 (2, 4, 6 and 8 wt%) are shown in Figure 5.2. Pure poly(vinyl alcohol) (PVA) is known to exhibit semi-crystalline character, commonly showing broad crystalline reflections near $2\theta \approx 20°$ and a weaker feature near $2\theta \approx 40°$, which are attributed to ordered PVA domains [6,7]. Upon blending PVA with poly(vinyl pyrrolidone) (PVP) the intensity of these crystalline reflections is reduced and the feature at ~40° seems to disappear, while the broad halo near ~20° becomes more pronounced — a signature of increased amorphicity caused by polymer-polymer miscibility and disruption of long-range crystalline order [8,9].

Addition of sodium sulfate further modifies the diffraction profile. With low salt concentration the PVA/PVP halo remains dominant; as Na_2SO_4 concentration is increased, new broad features or humps may appear at ≈30–33° and at higher angles (~40°), indicating the presence of salt-derived short-range structural correlations rather than well-developed salt crystals. These broad, low-intensity features are consistent with ionic clusters or polymer–salt complexes that have local order but lack the periodicity of bulk crystalline Na_2SO_4.

The progressive disappearance and broadening of crystalline peaks with increasing Na_2SO_4 content indicate that the salt disrupts PVA's semi-crystalline domains and increases the amorphous fraction of the blend. This effect is usually attributed to (i) ion–polymer complexation (coordination of Na^+ with PVA hydroxyls and interaction of SO_4^{2-} with polymer backbone polar groups), (ii) plasticization / salting-out effects that change hydrogen-bonding networks, and (iii) generation of microstrain and chain-segment displacement that broaden XRD features [10-12].

Mechanistically, the coordination of Na^+ ions to –OH (PVA) and C=O (PVP) functionalities weaken interchain crystalline interactions and promotes more disordered chain conformations. This local disruption produces microstrain and short-range disorder, which are observed as peak broadening and reduced crystalline intensity in XRD (rather than the appearance of sharp diffraction peaks that would indicate macroscopic salt crystallization). Consequently, the XRD evolution with increasing Na_2SO_4 supports the transformation from a semi-crystalline polymer host toward a more amorphous, ion-doped polymer matrix — a behavior reported for many PVA-based polymer electrolytes and composite membranes. [13]

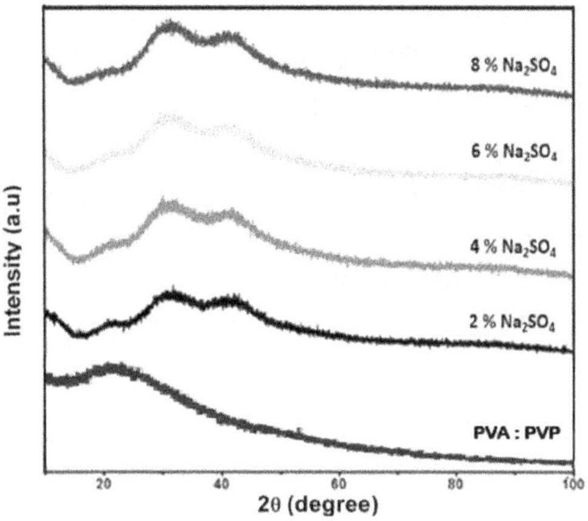

Figure 5.2 XRD patterns of PVA (80 wt.%): PVP (20 wt.%) and its blends with different concentrations of Na_2SO_4 (2, 4, 6 and 8 wt.%).

5.3.2 Fourier Transform Infrared (FTIR) Spectroscopy

FTIR spectroscopy was performed to study the molecular relations and confirm the complexation behaviour within the PVA/PVP/Na_2SO_4 solid polymer electrolyte system. Variations in peak position, shape, and intensity provide clear evidence of modifications in the polymer structure upon salt incorporation. The FTIR spectra of PVA, PVP, and their Na_2SO_4 -based complexes are illustrated in Figure 5.3.

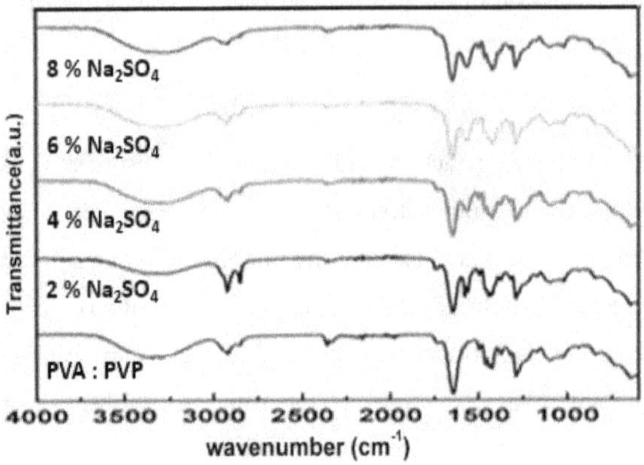

Figure 5.3 FTIR spectra of PVA (80 wt.%): PVP (20 wt.%) and its blends with different concentrations of Na_2SO_4 (2, 4, 6 and 8 wt.%).

The FTIR spectra of the individual polymers and the PVA:PVP:Na_2SO_4 blend provide clear evidence of molecular interactions induced by salt incorporation. Pure PVP exhibits a characteristic band near 1564 cm^{-1}, corresponding to the C=N stretching vibration of the pyrrolidone ring, and a band around 928 cm^{-1} assigned to out-of-plane C–H bending of the heterocyclic ring [8,9]. Pure PVA shows a broad absorption around 3340 cm^{-1}, attributed to O–H stretching of hydrogen-bonded hydroxyl groups, consistent with its semi-crystalline structure [14,15].

In the PVA:PVP:Na_2SO_4 blend, the principal vibrational bands at 3340, 2931, 1640, 1417, 1019, and 928 cm^{-1} correspond respectively to O–H stretching, CH_2 asymmetric stretching, C=O stretching, CH_2 bending, C–O stretching, and C–O symmetric stretching, in agreement with characteristic features reported for PVA–PVP systems [13]. The bands at 1019 and 928 cm^{-1} are associated with the *syndiotactic* sequences of PVA [16].

Upon incorporation of Na_2SO_4, systematic shifts in peak positions, broadening, and intensity changes occur—particularly in the O–H and C=O stretching regions—indicating ion–polymer complexation, where Na^+ interacts with hydroxyl groups of PVA and carbonyl groups of PVP [10,11][17]. These interactions weaken interchain hydrogen bonding and disturb local ordering, which is consistent with reports on alkali-salt-doped PVA and PVP polymer electrolytes [12,13].

The region 1640–1649 cm^{-1}, arising from C=O and C–O bending in PVP and PVA, exhibits subtle shifts and intensity reduction upon salt addition, further confirming the formation of polymer–salt complexes [17]. Increasing Na_2SO_4 concentration leads to an overall increase in amorphous character, consistent with analogous salt-doped PVA-based polymer electrolytes reported in the literature [10][13].

Overall, the FTIR evolution across compositions (Table 5.1) indicates that Na_2SO_4 is effectively complexed with the PVA: PVP host network, disrupting long-range order and enhancing the amorphous matrix—an essential condition for improved ionic transport in solid polymer electrolytes [10-13][17]

Table 5.1 Vibrational frequencies of PVA:PVP:Na_2SO_4 system.

Assignment	Wavenumber(cm-1)				
	PVA / PVP	**PVA/PVP/ 2% Na_2SO_4**	**PVA/PVP/ 4% Na_2SO_4**	**PVA/PVP/ 6% Na_2SO_4**	**PVA/PVP/ 8% Na_2SO_4**
O-H stretching	3340	3329	3314	3316	3311
CH2 asymmetric stretching	2931	2920	2921	2921	2920
C-H stretching	2843	2849	2849	2848	2851
CH2 bending	1417	1413	1419	1418	1419
C-O stretching	1019	1013	1021	1017	1015
C=O stretching	1640	1645	1648	1645	1649
C-N stretching	1564	1579	1569	1579	1564
C-O Symmetric stretching	928	932	934	930	930

5.3.3 Electrical Impedance

Impedance spectroscopy was employed to investigate the electrical response and microstructural features of the PVA/PVP– Na_2SO_4 polymer electrolyte films. The resulting Cole–Cole plots revealed the presence of depressed semicircular arcs, characteristic of polymer electrolytes exhibiting non-Debye relaxation behaviour [18]. Such semicircles arise from the combined contribution of bulk resistance and capacitance, which can be represented by an equivalent circuit model consisting of parallel resistance and capacitance elements. This behaviour indicates that ion transport within the polymer matrix is influenced not only by the bulk resistance of the electrolyte but also by interfacial polarization phenomena, which contribute to the overall impedance response of the system.

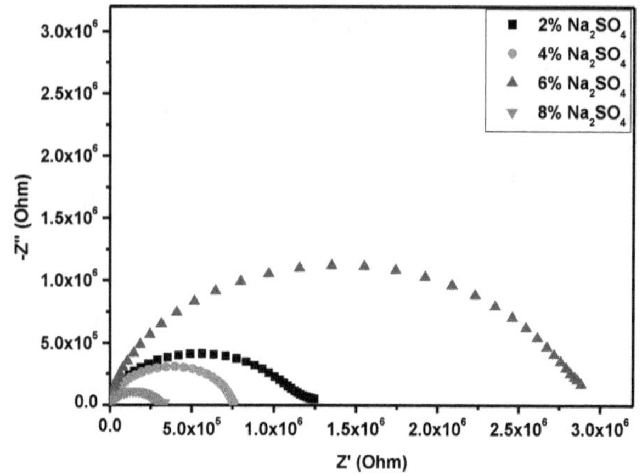

Figure 5.4 Nyquist plots obtained by ac impedance for different concentration of Na_2SO_4 (2, 4, 6 and 8 wt.%)-doped PVA/PVP SPEs at room temperature.

Among all compositions studied, the polymer film containing 8 wt.% Na_2SO_4showed the smallest semicircle diameter, corresponding to the lowest bulk resistance. This directly reflects enhanced ion mobility and effective charge transport pathways in the polymer matrix at this salt concentration.

The bulk resistance (R_b) was determined from the intercept of the semicircle with the real axis (Z′) of the Nyquist plot. Using this resistance value, the ionic conductivity (σ) of the films was calculated using equation no. 24. From this analysis, the maximum ionic conductivity was obtained for the 8 wt.% Na_2SO_4 doped PVA/PVP film, yielding a value of 2.4×10^{-7} S·cm^{-1} at room temperature. The enhanced conductivity at this composition is attributed to the increased number of dissociated Na^+ ions, which facilitate ion migration through the amorphous regions of the polymer blend.

5.3.4 Measurements of Conductance Spectra

The electrical conductivity spectra of Na_2SO_4 doped PVA/PVP polymer electrolytes as a function of frequency are presented in Figure 5.5. Across all compositions, the plots exhibit two distinct regions: a low-frequency plateau and a high-frequency dispersion zone. The low-frequency plateau corresponds to the dc conductivity (σdc), which arises from long-range ion migration within the polymer matrix. At higher frequencies, the conductivity increases progressively due to localized ion motion and relaxation phenomena, indicative of short-range hopping of charge carriers between available sites.

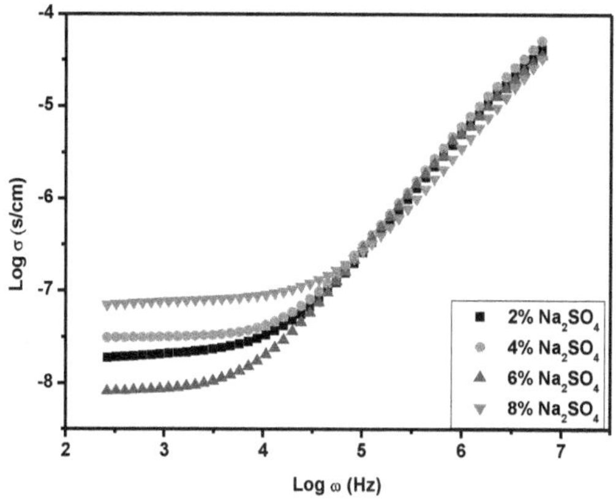

Figure 5.5 Frequency dependent conductivity of Na_2SO_4 doped PVA-PVP polymer sample.

Among the investigated samples, the composition containing 8 wt.% Na_2SO_4 demonstrated the highest σ_{dc}, confirming its superior ion transport capability. The enhanced conductivity at this concentration can be attributed to the optimal balance between salt dissociation and polymer chain flexibility, leading to a larger number of mobile Na^+ ions in the amorphous regions of the polymer blend. This frequency-dependent behaviour is consistent with Jonscher's universal power law, often observed in ion-conducting polymer systems, where the high-frequency dispersion is associated with correlated ion hopping and dielectric relaxation processes [19]. The results thus reaffirm that 8 wt.% Na_2SO_4-doped PVA/PVP electrolyte is the most efficient composition for sodium-ion transport, aligning with the impedance results discussed.

5.3.5 Frequency-Dependent Electric Modulus

The electric modulus spectra of Na_2SO_4-doped PVA/PVP polymer electrolytes at different salt concentrations are presented in Fig. 5.6. This analysis provides deeper insights into the relaxation dynamics of the system. Unlike dielectric constant studies, which are dominated at low frequencies by electrode polarization and space-charge effects due to ion accumulation at the electrode–electrolyte interface, the modulus formalism effectively minimizes such contributions, thereby emphasizing the bulk relaxation behaviour.

The spectra clearly exhibit distinct relaxation peaks whose positions and magnitudes vary with Na_2SO_4 concentration. At lower frequencies, the imaginary component of the modulus (M'') rises gradually and reaches a well-defined maximum before declining at higher frequencies. The observed relaxation peaks correspond to the ionic relaxation processes associated with the motion of Na^+ ions in the amorphous domains of the polymer matrix. The shift in peak position and intensity with increasing salt concentration indicates the influence of ion–polymer interactions and ion mobility on the relaxation dynamics.

Thus, the frequency-dependent modulus representation confirms that the relaxation behaviour in Na_2SO_4–PVA/PVP electrolytes is dominated by short-range ion hopping and polymer segmental dynamics rather than electrode polarization. The variations with salt concentration provide evidence of the optimal doping level required for achieving efficient ionic conduction in the blend system.

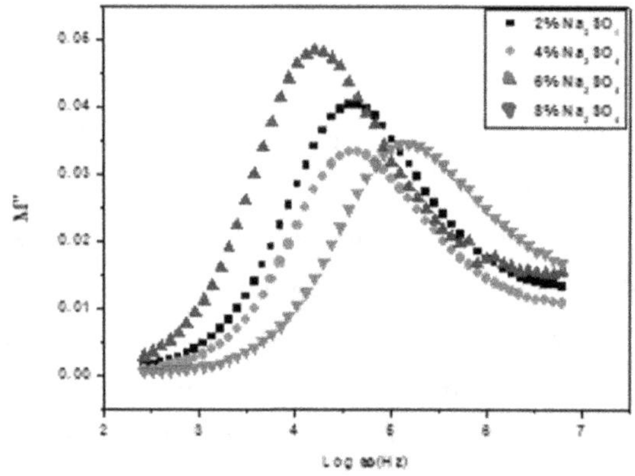

Figure 5.6 Frequency dependent electric modulus of Na_2SO_4 doped PVA-PVP polymer sample.

5.4 Conclusion

Electrochemical impedance studies on PVA–PVP–Na_2SO_4 solid polymer electrolytes provided a comprehensive understanding of ion dynamics and relaxation behaviour as a function of salt concentration. The impedance spectra confirmed that the bulk response of the system is governed by ionic conduction within the amorphous regions of the polymer, while electrode polarization effects are effectively suppressed under modulus formalism. The observed frequency-dependent relaxation peaks highlight the coupling between Na^+ ion mobility and the segmental motion of the polymer chains.

Among the investigated compositions, the 8 wt.% Na_2SO_4-doped PVA/PVP film exhibited the most favourable characteristics, delivering the highest conductivity value of 2.4×10^{-7} S·cm^{-1} at room temperature. This optimum composition demonstrated efficient ion hopping and good polymer–ion coordination.

This study establishes that the Na_2SO_4-incorporated PVA/PVP blend electrolyte achieves enhanced ionic transport through short-range ion hopping assisted by polymer chain flexibility, with the 8 wt.% composition offering the most effective balance for potential application in sodium-based solid-state electrochemical devices. When compared to LiI & $LiNO_3$-doped PVDF-HFP/PVP systems, the conductivity of the present Na_2SO_4–PVA/PVP electrolyte is relatively lower; however, its environmentally friendly, material abundance, and cost-effectiveness make it a more sustainable choice for sodium-based energy storage applications.

5.5 References

[1] Gogoi, B.J.; Murugesan, M.; Nallamuthu, N.; Devendran, P.; Murugan, A.; Blange, R.; Shellaiah, M. Investigation into PVDF-HFP and PVP Polymer Blend Electrolytes with Lithium Ions for Energy Storage Application. *Polymers* **2025**, *17*, 1758.

[2] Gami, P.; Patel, D.; Sharma, R. Review on PVDF-HFP/PVP-Based Polymer Blend Electrolytes for Lithium-Ion Conduction. *Mater. Adv.* **2025**, *6*, 710–729.

[3] Saikia, D.; Kumar, A. Ionic Conduction in P(VDF-HFP)/PVDF–(PC + DEC)–$LiClO_4$ Polymer Gel Electrolytes. *Electrochim. Acta* **2004,** *49*, 2581–2589.

[4] Dennis, J.O.; Ahmed, A.Y.; Khir, M.H.M. Polyvinyl Alcohol-Based Solid Polymer Electrolytes: A Review on Ionic Conduction and Applications in Energy Storage Devices. *Molecules* **2023**, *28*, 1781.

[5] Vareda, J.P.; Correia, J.R.; Durães, L. Silica–PVA Aerogels with Sodium Perchlorate as Potential Solid Polymer Electrolytes. *Gels* **2024**, *10*, 293.

[6] Brza, M.A.; B. Aziz, S.; Anuar, H.; Dannoun, E.M.A.; Ali, F.; Abdulwahid, R.T.; Al-Zangana, S.; Kadir, M.F.Z. The Study of EDLC Device with High Electrochemical Performance Fabricated from Proton Ion Conducting PVA-Based Polymer Composite Electrolytes Plasticized with Glycerol. *Polymers* **2020**, *12*, 1896.

[7] Aree, T.; Chauhan, C.; Kumar, S.; Kumar, R. Salient Optical and Electrical Properties of Poly(vinyl alcohol) Films Inculcated by Two Copper(II) Complexes. *J. Phys. Chem. C* **2025**, *129*(33), 15131–15150.

[8] Sheik, S.; Nagaraja, G. K.; Prashantha, K. Effect of Silk Fiber on the Structural, Thermal, and Mechanical Properties of PVA/PVP Composite Films. *Polymer Engineering and Science* **2017**, *57* (12), 1321–1330.

[9] Morad, M. A.; Abo Ghazala, M. S.; El-Shaarawy, M. G.; Gouda, M. E.; Elrasasi, T. Y. Preparation and Characterization of Conjugated PVA/PANi Blend Films Doped with Functionalized Graphene for Thermoelectric Applications. *Sci. Rep.* **2024**, *14*, 16722.

[10] Bhargav, P. B.; Mohan, V. M.; Sharma, A. K.; Rao, V. V. R. N. Investigations on Electrical Properties of (PVA:NaF) Polymer Electrolytes for Electrochemical Cell Applications. *Curr. Appl. Phys.* **2009**, *9* (1), 165–171.

[11] Shenbagavalli, S.; Muthuvinayagam, M.; Revathy, M. S. Enhancement of Electrical and Electrochemical Properties of Sodium Bromide Incorporated with Poly(ethylene oxide)/Poly(vinylidene fluoride-hexafluoropropylene) Solid Blend Polymer Electrolytes for Electrochemical Double Layer Capacitors. *J. Energy Storage* **2022**, *55*, 105726.

[12] Srivastava, S.; Varshney, P. K. A Structural Study of Mixed Ion PVA-Based Composite Polymer Electrolyte Using X-ray Diffraction Studies. *Int. J. Appl. Eng. Res.* **2017**, *12* (11), 2926–2928.

[13] Hadi, J. M.; Aziz, B. K.; Aziz, S. B. Structural and Electrochemical Properties of Ternary Solid Polymer Electrolytes Based on PVA:CS:FSG Doped with Sodium Thiocyanate. *Sci. Rep.* **2025**, *15*, 14379.

[14] Colthup, N. B.; Daly, L. H.; Wiberley, S. E. *Introduction to Infrared and Raman Spectroscopy*; Academic Press: San Diego, CA, 1990.

[15] Socrates, G. *Infrared and Raman Characteristic Group Frequencies: Tables and Charts*, 3rd ed.; Wiley, 2004.

[16] Nasar, G.; Khan, M. S.; Khalil, U. Structural Study of PVA Composites with Inorganic Salts by X-ray Diffraction. *J. Basic Appl. Sci.* **2009**, *5*, 1–6.

[17] Aziz, S. B.; Woo, T. J.; Kadir, M. F. Z.; Ahmed, H. M. A Conceptual Review on Polymer Electrolytes and Ion Transport Models. *J. Sci. Adv. Mater. Devices* **2018**, *3*, 1–17.

[18] Srivastava, N.; Varshney, P.K. Conductivity and Structural Studies of PVA-Based Mixed-Ion Composite Polymer Electrolytes. *Ionics* **2019**, *25*, 5011–5022.

[19] Jonscher, A.K. The "Universal" Dielectric Response. *Nature* **1977**, *267*, 673–679.